Electric and Hybrid Vehicles: Principles, Design and Technology

Electric and Hybrid Vehicles: Principles, Design and Technology

Mary Murphy

Larsen & Keller
www.larsen-keller.com

Electric and Hybrid Vehicles: Principles, Design and Technology
Mary Murphy
ISBN: 978-1-64172-127-1 (Hardback)

Larsen & Keller
Published by Larsen and Keller Education,
5 Penn Plaza,
19th Floor,
New York, NY 10001, USA

Cataloging-in-Publication Data

Electric and hybrid vehicles : principles, design and technology / Mary Murphy.
p. cm.
Includes bibliographical references and index.
ISBN 978-1-64172-127-1
1. Electric vehicles. 2. Hybrid electric vehicles. 3. Electric automobiles. I. Murphy, Mary.
TL220 .E44 2019
629.229 3--dc23

For more information regarding Larsen and Keller Education and its products, please visit the publisher's website www.larsen-keller.com

Table of Contents

Preface

An electric vehicle is a vehicle that uses a single or multiple electric or traction motors for propulsion. It may be powered by electricity derived from off-vehicle sources, or from a self-contained system of solar panels, battery or an electric generator. A hybrid electric vehicle (HEV) is a kind of hybrid vehicle, which integrates an electric propulsion system with an internal combustion engine. There are different types of HEV, of which the hybrid electric car is the most common form. Depending on the way in which power is delivered to the drivetrain, HEV can be classified into parallel hybrids, series hybrids and power-split hybrids. They can also be classified into mild and full hybrids depending on the degree of hybridization. Modern HEVs use innovative technologies such as regenerative brakes, motor-generators and start-stop systems, besides others. This book unfolds the innovative aspects of electric and hybrid vehicles. Most of the topics introduced in this book cover the principles, design and technology of HEV. As this field is emerging at a rapid pace, the contents of this book will help the readers understand the modern concepts and trends in this field.

A foreword of all Chapters of the book is provided below:

Chapter 1, Electric vehicle are vehicles that use traction motors or electric motors for propulsion. Such vehicles may be powered by an inbuilt battery or through a collector system with electricity derived from off-vehicle sources. This chapter provides an introduction to electric vehicles and provides an overview of different electric vehicles such as electric car, electric bike and electric bus; **Chapter 2**, Hybrid vehicles use more than one distinct power types. These can be two-wheeled and cycle-type vehicles or heavy vehicles. The aim of this chapter is to provide an understanding of hybrid electric vehicles, such as hybrid electric vehicle, plug-in hybrid electric vehicle, hybrid electric truck and hybrid electric bus, besides others; **Chapter 3**, A fuel cell electric vehicle is a kind of electric vehicle that uses a fuel cell in combination with a battery or a supercapacitor for powering its electric motor. The topics elaborated in this chapter on fuel cell, fuel cell vehicle, fuel cell forklift and fuel cell electric bus will help in providing a better perspective about the different types of fuel cell electric vehicles and their functioning; **Chapter 4**, Electric vehicles are powered using one or more electric motors. It may contain solar panels, a battery or an electric generator for converting fuel into electricity. This chapter provides a close examination of electric vehicle technologies and includes vital topics like electric vehicle battery, charge control, electric motor, traction motor, electric traction system, motor controller, etc. for a comprehensive understanding of the subject; **Chapter 5**, An electric vehicle charging station is an infrastructure that supplies electric power for recharging electric vehicles, such as plug-in hybrids, electric cars, etc. This chapter has been carefully written to provide an easy understanding of the varied aspects of electric vehicle charging devices and systems, with a compilation of significant topics such as charging station, connectors, rapid chargers, fast chargers, slow chargers, wireless charging, EnergyBus connectors, etc.

I would like to thank the entire editorial team who made sincere efforts for this book and my family who supported me in my efforts of working on this book. I take this opportunity to thank all those who have been a guiding force throughout my life.

Mary Murphy

1

Introduction to Electric Vehicles

Electric vehicle are vehicles that use traction motors or electric motors for propulsion. Such vehicles may be powered by an inbuilt battery or through a collector system with electricity derived from off-vehicle sources. This chapter provides an introduction to electric vehicles and provides an overview of different electric vehicles such as electric car, electric bike and electric bus.

An electric vehicle (EV), also referred to as an electric drive vehicle, is a vehicle which uses one or more electric motors for propulsion. Depending on the type of vehicle, motion may be provided by wheels or propellers driven by rotary motors, or in the case of tracked vehicles, by linear motors. Electric vehicles can include electric cars, electric trains, electric trucks, electric lorries, electric airplanes, electric boats, electric motorcycles and scooters, and electric spacecraft.

Electric vehicles are different from fossil fuel-powered vehicles in that they can receive their power from a wide range of sources, including fossil fuels, nuclear power, and renewable sources such as tidal power, solar power, and wind power or any combination of those. However it is generated, this energy is then transmitted to the vehicle through use of overhead lines, wireless energy transfer such as inductive charging, or a direct connection through an electrical cable. The electricity may then be stored onboard the vehicle using a battery, flywheel, super capacitor, or fuel cell. Vehicles making use of engines working on the principle of combustion can usually only derive their energy from a single or a few sources, usually non-renewable fossil fuels. A key advantage of electric or hybrid electric vehicles is their ability to recover braking energy as electricity to be restored to the on-board battery (regenerative braking) or sent back to the grid (V2G). At the beginning of the 21st century, increased concern over the environmental impact of the petroleum-based transportation infrastructure, along with the spectra of peak oil, led to renewed interest in an electric transportation infrastructure. As such, vehicles which can potentially be powered by renewable energy sources, such as hybrid electric vehicles or pure electric vehicles, are becoming more popular.

In an electric vehicle (EV), a battery or other energy storage device is used to store the electricity that powers the motor. EV batteries must be replenished by plugging in the vehicle to a power source. Some electric vehicles have onboard chargers; others plug into a charger located outside the vehicle. Both types, however, use electricity that comes from the power grid. Although electricity production may contribute to air pollution, EVs are considered zero-emission vehicles because their motors produce no exhaust or emissions.

There are currently no light-duty electric vehicles available from the major auto manufacturers. Neighborhood electric vehicles (NEVs), on the other hand, are being manufactured by a variety of companies. These small vehicles are commonly used for neighborhood commuting, light hauling, and delivery. Their use is limited to areas with 35 mph speed limits or for off-road service on college campus or at airports or resort areas.

Because they are limited to speeds of 25 mpg or less, NEVs are not considered light-duty vehicles and are not eligible for fleet credit under the Energy Policy Act of 1992 Standard Compliance option and Federal Fleet Requirements. However, their versatility in moving people through limited commute areas makes them useful in a variety of applications. Other useful EVs in niche applications include electric scooters and bikes.

Carbon Dioxide Emissions

An EV recharged from the existing US grid electricity emits about 115 grams of CO_2 per kilometer driven (6.5 oz(CO_2)/mi), whereas a conventional US-market gasoline powered car emits 250 g(CO_2)/km (14 oz(CO_2)/mi) (most from its tailpipe, some from the production and distribution of gasoline). The savings are questionable relative to hybrid or diesel cars, (according to official British government testing the most efficient European market cars are well below 115 grams of CO_2 per kilometer driven), but would be more significant in countries with cleaner electric infrastructure. In a worst case scenario where incremental electricity demand would be met exclusively with coal, a 2009 study conducted by the WWF, World Wildlife Foundation, and IZES found that a mid-size EV would emit roughly 200 g(CO_2)/km (11 oz(CO_2)/mi), compared with an average of 170 g(CO_2)/km (9.7 oz(CO_2)/mi) for a gasoline powered compact car. This study concluded that introducing 1 million EV cars to Germany would, in the best case scenario, only reduce CO_2 emissions by 0.1%, if nothing is done to upgrade the electricity infrastructure or manage demand.

Like any other vehicles, EVs themselves of course differ in their fuel efficiency and their total cost of ownership, including the environmental costs of their manufacture and disposal.

According to the US Department of Energy, most electricity generation in the United States is from fossil sources, and half of that is from coal. Coal is more carbon-intensive than oil. Overall average efficiency from US power plants (33% efficient) to point of use (transmission loss 9.5%) is 30%. Accepting a 70% to 80% efficiency for the electric vehicle gives a figure of only around 20% overall efficiency when recharged from fossil fuels. That is comparable to the efficiency of an internal combustion engine running at variable load. The efficiency of a gasoline engine is about 16%, and 20% for a diesel engine. This is much lower than the efficiency when running at constant load and optimal rotational speed, which gives efficiency around 30% and 45% respectively. The electric battery suffers from similar decrease in efficiency when running at variable load, which accounts for the modest increased efficiency of hybrid vehicles. The actual result in terms of emissions depends on different refining and transportation costs getting fuel to a car versus a power plant. Diesel engines can also easily run on renewable fuels, biodiesel, vegetable oil fuel, with no loss of efficiency. Using fossil based grid electricity substantially negates the in-vehicle efficiency advantages of electric cars. The major potential benefit of electric cars is to allow diverse renewable electricity sources to fuel cars.

According to the US Department of Energy, CO_2 emissions for electricity generated from coal result in 2.05 lb (0.93 kg) of CO_2 per kW·h or roughly 0.5 lb(CO_2)/mi (0.14 kg(CO_2)/km). CO_2 emissions from electricity produced from all types of fuel using the mix of sources in the US as of 2008 results in 1.35 lb (0.61 kg) of CO_2 per kW·h or 0.337 pounds of CO_2 per mile (0.095 kg(CO_2)/km) from an electric vehicle with a 0.250-kilowatt-hour-per-mile (0.155 kW·h/km; 0.56 MJ/km) energy consumption (typical). Gasoline used in Internal Combustion Engine automobiles produces 19.5 lbCO2/US gal (2.34 kg(CO_2)/L) directly and an undetermined amount of CO_2 in refining the crude oil, and transporting the gasoline to retail point of sale. With a US fleet average of 21.3 mpg_{-US} (11.0 L/100 km; 25.6 mpg_{-imp}) in 2008, this would indicate a CO_2 production of 0.915 lb/mi (0.258 kg/km) driven. Electric powered automobiles, even using the most CO_2 intensive coal produced electricity, produce half the emissions of gasoline powered automobiles.

If solar, wind, hydro, or nuclear electric generation, or carbon capture for fossil fuel powered plants were to become prevalent, electric vehicles could produce less CO_2, potentially zero. Based on GREET simulations, electric cars can achieve up to 100% reductions with renewable electric generation, against 77% with a B100 car. At present only a 32% reduction of CO_2 is available for electric cars recharging from non-renewable utilities on the US Grid, because of the majority use of fossil fuels in generation, and inefficiency in the grid itself.

Acceleration and Drivetrain Design

Electric motors can provide high power to weight ratios, and batteries can be designed to supply the large currents to support these motors.

Although some electric vehicles have very small motors, 15 kW (20 hp) or less and therefore have modest acceleration, many electric cars have large motors and brisk acceleration. In addition, the relatively constant torque of an electric motor, even at very low speeds tends to increase the acceleration performance of an electric vehicle relative to that of the same rated motor power internal combustion engine. Another early solution was American Motors' experimental Amitron piggyback system of batteries with one type designed for sustained speeds while a different set boosted acceleration when needed.

Electric vehicles can also use a direct motor-to-wheel configuration which increases the amount of available power. Having multiple motors connected directly to the wheels allows for each of the wheels to be used for both propulsion and as braking systems, thereby increasing traction. In some cases, the motor can be housed directly in the wheel, such as in the Whispering Wheel design, which lowers the vehicle's center of gravity and reduces the number of moving parts. When not fitted with an axle, differential, or transmission, electric vehicles have less drivetrain rotational inertia.

When the foot is lifted from the accelerator of an ICE, engine braking causes the car to slow. An EV would coast under these conditions, and applying mild regenerative braking instead provides a more familiar response.

A gearless or single gear design in some EVs eliminates the need for gear shifting, giving such vehicles both smoother acceleration and smoother braking. Because the torque of an electric motor is a function of current, not rotational speed, electric vehicles have a high torque over a larger range of speeds during acceleration, as compared to an internal combustion engine. As there is no delay in developing torque in an EV, EV drivers report generally high satisfaction with acceleration.

For example, the Venturi Fetish delivers supercar acceleration despite a relatively modest 220 kW (295 hp), and top speed of around 160 km/h (100 mph). Some DC motor-equipped drag racer EVs, have simple two-speed transmissions to improve top speed. The Tesla Roadster prototype can reach 100 km/h (62 mph) in 4 seconds with a motor rated at 185 kW (248 hp).

Vehicle Types

All kinds of vehicles can be modified by equipping them with some form of electric powertrain. Accordingly, we can have many different types of vehicles.

Hybrid Electric Vehicle

A hybrid electric vehicle combines a conventional (usually fossil fuel-powered) powertrain with some form of electric propulsion. Common examples include hybrid electric cars such as the Toyota Prius.

On- and off-road Electric Vehicles

Electric vehicles are on the road in many functions, including electric cars, electric trolleybuses, electric bicycles, electric motorcycles and scooters, neighborhood electric vehicles, golf carts, milk floats, and forklifts. Off-road vehicles include electrified all-terrain vehicles and tractors.

Railborne Electric Vehicles

The fixed nature of a rail line makes it relatively easy to power electric vehicles through permanent overhead lines or electrified third rails, eliminating the need for heavy onboard batteries. Electric locomotives, electric trams/streetcars/trolleys, electric light rail systems, and electric rapid transit are all in common use today, especially in Europe and Asia.

Since electric trains do not need to carry a heavy internal combustion engine or large batteries, they can have very good power-to-weight ratios. This allows high speed trains such as France's double-deck TGVs to operate at speeds of 320 km/h (200 mph) or higher, and electric locomotives to have a much higher power output than diesel locomotives. In addition they have higher short-term surge power for fast acceleration, and using regenerative braking can put braking power back into the electrical grid rather than wasting it.

Maglev trains are also nearly always electric vehicles.

Airborne Electric Vehicles

Since the beginning of the era of aviation, electric power for aircraft has received a great deal of experimentation. Currently flying electric aircraft include manned and unmanned aerial vehicles.

Seaborne Electric Vehicles

Electric boats were popular around the turn of the 20th century. Interest in quiet and potentially renewable marine transportation has steadily increased since the late 20th century, as solar cells have given motorboats the infinite range of sailboats. Submarines use batteries (charged by diesel or gasoline engines at the surface), nuclear power, or fuel cells run electric motor driven propellers.

Spaceborne Electric Vehicles

Electric power has a long history of use in spacecraft. The power sources used for spacecraft are batteries, solar panels and nuclear power. Current methods of propelling a spacecraft with electricity include the arcjet rocket, the electrostatic ion thruster, the Hall effect thruster, and Field Emission Electric Propulsion. A number of other methods have been proposed, with varying levels of feasibility.

Online Electric Vehicle

OLEV Bus is an electric vehicle that can be charged while stationary or driving.

The Online Electric Vehicle (OLEV) is an electric vehicle that can be charged while stationary or driving, thus removing the need to stop at a charging station. Likewise, an OLEV tram does not require pantographs to feed power from electric wires strung above the tram route.

OLEV is a groundbreaking technology that accelerates the development of purely electric vehicles as a viable option for future transportation systems, be they personal vehicles or public transit. This is accomplished by solving technological issues that limit the commercialization of electric vehicles such as price, weight, volume, driving distance, and lack of charging infrastructure.

OLEV receives power wirelessly through the application of the "Shaped Magnetic Field in Resonance (SMFIR)" technology. SMFIR is a new technology introduced by KAIST that enables electric vehicles to transfer electricity wirelessly from the road surface while moving. Power comes from the electrical cables buried under the surface of the road, creating magnetic fields. There is a receiving device installed on the underbody of the OLEV that converts these fields into electricity. The length of power strips installed under the road is generally 5%-15% of the entire road, requiring only a few sections of the road to be rebuilt with the embedded cables.

Two OLEV buses will run an inner city route in Gumi, Republic of Korea

OLEV has a small battery (one-third of the size of the battery equipped with a regular electric car). The vehicle complies with the international electromagnetic fields (EMF) standards of 62.5 mG, within the margin of safety level necessary for human health. The road has a smart function as well, to distinguish OLEV buses from regular cars—the segment technology is employed to control the power supply by switching on the power strip when OLEV buses pass along, but switching it off for other vehicles, thereby preventing EMF exposure and standby power consumption. As of today, the SMFIR technology supplies 60 kHz and 180 kW of power remotely to transport vehicles at a stable, constant rate.

Mechanical Description

The OLEV system is split into two main parts: the segmented "recharging" road and the "pick-up" modules on the vehicle.

In the Road

In the "recharging" road, slim W-shaped ferrite cores (magnetic cores used in induction) are buried 30 cm underground in a fish bone like structure. Power cables are wrapped around the center of the fish bone structures to make the "primary coils". This design combines the magnetic fields of the two sides of the cables and shapes the fields in a way that maximizes induction. Moreover, the primary coils are placed in segments across certain spans of the road so that only about 5% to 15% of the road needs to be remodeled. To power the primary coils, the cables are attached to the South Korean national power grid through a power inverter. The inverter accepts 60Hz 3-phase 380 or 440 voltage from the grid to generate 20kHz of AC electricity into the cables. In turn, the cables create a 20 kHz magnetic field that sends flux through the slim ferrite cores to the pick-ups on the OLEV.

On the OLEV

Attached beneath the vehicle, are "pick-up" modules, or the secondary coils, that consist of wide W-shaped ferrite cores with wires wrapped around the center. When the pick-ups "pick up" the flux from the primary coils, each pick-up gains about 17 kW of power from the induced current. This power is sent to the electric motor and battery through a regulator (a managing device that can distribute power based on need), thereby charging the OLEV wirelessly.

Models

Model	Weight	Core Shape in Primary Coil	Core Shape in Secondary Coil	Air Gap Between Road and Pick-up	Power Efficiency	Power Gained per Pick-up	Electrical Horse Power	Current in Primary Coil	Additional Mech-anism
Generation 1 (Small Cart)	10 kg	E shape	E shape	1 cm	80%	3 kW	4.02Hp	100 Amp	Vertical Alignment Mechanism for 3mm
Generation 2 (Bus)	80 kg	U shape	Long, flat	17 cm	72%	6 kW	8.04Hp	200 Amp	Return cables for primary coils
Generation 3 (SUV)	110 kg	Slim W shape	Wide W shape	17 cm	71%	17 kW	22.79Hp	200 Amp	None

As seen in the table above, the generation 1 OLEV lacks a realistic margin for error. The lower current means a smaller magnetic field and requires the secondary coil to be very close to the floor, which can be an issue while driving. Moreover, if the primary and secondary coils are vertically misaligned by a distance over 3mm, the power efficiency drops greatly.

To fix these issues, KAIST came up with the generation 2 OLEV. In the gen 2 OLEV, the current in the primary coil was doubled to create a stronger magnetic field that allows for a larger air gap. The ferrite cores in the primary coils were changed to a U shape and the cores in the secondary coil were changed a flat board shape to pick-up as much flux as possible. This design allows the vertical misalignment to be about 20 cm with a 50% power efficiency. However, the U shaped cores also require return cables which bumps up the cost of production. Overall, the gen 2 made up for the gen 1's margins but, was more costly.

In response to the cost issue of gen 2, the third generation OLEV was developed. The third generation OLEV, uses ultra-slim W-shaped ferrite cores in the primary coil to reduce the amount of ferrite used to 1/5 of gen 2 and to remove the need of return cables. The secondary coil uses a thicker variation of the w-shaped cores as a way to make up for the lesser area for the magnetic flux to flow through compared to gen 2. Overall, the gen 3 OLEV made up for the gen 1's small margins and gen 2's increased cost.

Benefits & Issues

Benefits

- Zero emissions;
- 31% operating costs compared to gas powered counterparts;
- Lower maintenance and manufacturing costs;
- No charging station necessary;
- Can be stored like regular vehicles.

Issues

- Modern power grids cannot handle OLEV's on a large scale;

- Installation is costly;
- Can run out of power in heavy traffic;
- 40 mph speed limit;
- Cannot operate during power outages.

Plug-in Electric Vehicle

The Nissan Leaf (left) is the world's all-time top-selling highway-legal all-electric car (over 300,000), and the Chevrolet Volt (right) is the world's best-selling plug-in hybrid

A plug-in electric vehicle (PEV) is any motor vehicle that can be recharged from an external source of electricity, such as wall sockets, and the electricity stored in the rechargeable battery packs drives or contributes to drive the wheels. PEV is a subset of electric vehicles that includes all-electric or battery electric vehicles (BEVs), plug-in hybrid vehicles (PHEVs), and electric vehicle conversions of hybrid electric vehicles and conventional internal combustion engine vehicles. In China, plug-in electric vehicles are called new energy vehicles (NEVs).

Plug-in cars have several benefits compared to conventional internal combustion engine vehicles. They have lower operating and maintenance costs, and produce little or no local air pollution. They reduce dependence on petroleum and may severally reduce greenhouse gas emissions, depending on the electricity source, as motors are typically much more efficient than their engine equivalents. Plug-in hybrids capture most of these benefits when they are operating in all-electric mode. Despite their potential benefits the stock of plug-in electric cars represented just 0.15% of the 1.4 billion motor vehicles on the world's roads by the end of 2016.

BYD sold the most plug-in electric cars in 2015 and built more plug-in cars than any other maker in 2016.

As of December 2016, there were over 60 models of highway capable plug-in electric passenger cars and light-duty utility vans available for retail sales in the world. Cumulative global sales of highway legal plug-in electric passenger cars and light utility vehicles achieved the 2 million unit milestone in December 2016. The 3 million milestone was achieved in November 2017. The global ratio between the stock of all-electrics (BEVs) and plug-in hybrids (PHEVs) was 61:39 at the end of 2016. As of January 2018, the Nissan Leaf is the world's top selling highway-capable all-electric car in history, with global sales of over 300,000 units. The Tesla Model S ranks second with more than 200,000 units sold worldwide as of November 2017.

BYD ranks 1 In the world for plug-in electric car sales in 2017. As of December 2017, China has the world's largest stock of highway legal light-duty plug-in electric vehicles with cumulative sales of more than 1.2 million plug-in electric passenger cars. Among country markets, the United States ranks second with almost 765,000 plug-in electric cars sold through December 2017. More than 943,000 light-duty plug-in electric passenger cars have been registered in Europe up until December 2017, with sales in the light-duty plug-in electric segment led by Norway with over 200,000 units registered. China is the world's leader in the plug-in heavy-duty segment, including electric all-electric buses, and plug-in commercial and sanitation trucks, with a stock of new energy vehicles sold of more than 1,7 million units through December 2017. As of December 2016, China was the world's largest plug-in electric bus market with a stock of almost 343,500 vehicles.

Terminology

Brammo Empulse electric motorcycle at a charging station.

Plug-in Electric Vehicle

A plug-in electric vehicle (PEV) is any motor vehicle with rechargeable battery packs that can be charged from the electric grid, and the electricity stored on board drives or contributes to drive the wheels for propulsion. Plug-in electric vehicles are also sometimes referred to as grid-enabled vehicles (GEV) and also as electrically chargeable vehicles.

PEV is a subcategory of electric vehicles that includes battery electric vehicles (BEVs), plug-in hybrid vehicles, (PHEVs), and electric vehicle conversions of hybrid electric vehicles and conventional internal combustion engine vehicles. Even though conventional hybrid electric vehicles (HEVs) have a battery that is continually recharged with power from the internal combustion engine and regenerative braking, they can not be recharged from an off-vehicle electric energy source, and therefore, they do not belong to the category of plug-in electric vehicles.

"Plug-in electric drive vehicle" is the legal term used in U.S. federal legislation to designate the category of motor vehicles eligible for federal tax credits depending on battery size and their all-electric range. In some European countries, particularly in France, "electrically chargeable vehicle" is the formal term used to designate the vehicles eligible for these incentives. While the term "plug-in electric vehicle" most often refers to automobiles or "plug-in cars", there are several other types of plug-in electric vehicle, including electric motorcycles and scooters, neighborhood electric vehicles or microcars, city cars, vans, buses, electric trucks or lorries, and military vehicles.

Battery Electric Vehicles

U.S. Army GEM e2 neighborhood electric vehicle.

A battery electric vehicle (BEV) uses chemical energy stored in rechargeable battery packs as its only source for propulsion. BEVs use electric motors and motor controllers instead of internal combustion engines (ICEs) for propulsion.

A plug-in hybrid operates as an all-electric vehicle or BEV when operating in charge-depleting mode, but it switches to charge-sustaining mode after the battery has reached its minimum state of charge (SOC) threshold, exhausting the vehicle's all-electric range (AER).

Plug-in Hybrid Electric Vehicles

Several Prius+ converted plug-in hybrids.

A plug-in hybrid electric vehicle (PHEV or PHV), also known as a plug-in hybrid, is a hybrid electric vehicle with rechargeable batteries that can be restored to full charge by connecting a plug to an external electric power source. A plug-in hybrid shares the characteristics of both a conventional hybrid electric vehicle and an all-electric vehicle: it uses a gasoline engine and an electric motor for propulsion, but a PHEV has a larger battery pack that can be recharged, allowing operation in all-electric mode until the battery is depleted.

Aftermarket Conversions

An aftermarket electric vehicle conversion is the modification of a conventional internal combustion engine vehicle (ICEV) or hybrid electric vehicle (HEV) to electric propulsion, creating an all-electric or plug-in hybrid electric vehicle.

There are several companies in the U.S. offering conversions. The most common conversions have been from hybrid electric cars to plug-in hybrid, but due to the different technology used in hybrids by each carmaker, the easiest conversions are for 2004–2009 Toyota Prius and for the Ford Escape/Mercury Mariner Hybrid.

New Energy Vehicles

In China the term new energy vehicles (NEVs) refers to vehicles that are partially or fully powered by electricity, such as battery electric vehicles (BEVs) and plug-in hybrids (PHEVs). The Chinese government began implementation of its NEV program in 2009 to foster the development and introduction of new energy vehicles.

Advantages

Lower Operating and Maintenance Costs

Internal combustion engines are relatively inefficient at converting on-board fuel energy to propulsion as most of the energy is wasted as heat, and the rest while the engine is idling. Electric motors, on the other hand, are more efficient at converting stored energy into driving a vehicle. Electric drive vehicles do not consume energy while at rest or coasting, and modern plug-in cars can capture and reuse as much as one fifth of the energy normally lost during braking through regenerative braking. Typically, conventional gasoline engines effectively use only 15% of the fuel energy content to move the vehicle or to power accessories, and diesel engines can reach on-board efficiencies of 20%, while electric drive vehicles typically have on-board efficiencies of around 80%.

The operating cost of the Toyota Prius Plug-in Hybrid in the U.S. is estimated at US$0.03 per mile while operating in all-electric mode.

In the United States, as of early 2010 with a national average electricity rate of US$0.10 per kWh, the cost per mile for a plug-in electric vehicle operating in all-electric mode is estimated between $0.02 to $0.04, while the cost per mile of a standard automobile varies between $0.08 to $0.20, considering a gasoline price of $3.00 per gallon. As petroleum price is expected to increase in the future due to oil production decline and increases in global demand, the cost difference in favor of PEVs is expected to become even more advantageous.

According to Consumer Reports, as of December 2011 the Nissan Leaf has a cost of 3.5 cents per mile and the Chevrolet Volt has a cost in electric mode of 3.8 cents per mile. The Volt cost per mile is higher because it is heavier than the Leaf. These estimates are based on the fuel economy and energy consumption measured on their tests and using a U.S. national average rate of 11 cents per kWh of electricity. When the Volt runs in range-extended mode using its premium gasoline-powered engine, the plug-in hybrid has a cost of 12.5 cents per mile. The out-of-pocket cost per mile of the three most fuel efficient gasoline-powered cars as tested by the magazine are the Toyota Prius, with a cost of 8.6 cents per miles, the Honda Civic Hybrid with 9.5 cents per mile, the Toyota Corolla with 11.9 cents per mile, and the Hyundai Elantra 13.1 cents per mile. The analysis also found that on trips up to 100 mi (160 km), the Volt is cheaper to drive than the Prius and the other three cars due to the Volt's 35 mi (56 km) driving range on electricity. The previous operating costs do not include maintenance, depreciation or other costs.

All-electric and plug-in hybrid vehicles also have lower maintenance costs as compared to internal combustion vehicles, since electronic systems break down much less often than the mechanical systems in conventional vehicles, and the fewer mechanical systems on board last longer due to the better use of the electric engine. PEVs do not require oil changes and other routine maintenance checks.

The Edison Electric Institute (EEI) conducted an analysis that demonstrated that between January 1976 and February 2012 the real price for gasoline has been much more volatile than the real price of electricity in the United States. The analysis is based on a plug-in electric vehicle with an efficiency of 3.4 miles per kW-hr (like the Mitsubishi i MiEV) and a gasoline-powered vehicle with a fuel economy rated at 30 mpg_{US} (7.8 L/100 km; 36 mpg_{imp}) (like the 2012 Fiat 500). The EEI estimated that operating a plug-in would have had an equivalent cost of around US$1.50 a gallon in the late 1970s and early 1980s, and around US$1.00 a gallon since the late 1990s. In contrast, the price to operate an internal combustion engine vehicle has had much ample variations, costing more than US$3.50 per gallon during the 1979 energy crisis, then had a couple of lows with prices at less than US$1.50 during 1999 and 2001, only to climb and reach a maximum of more than US$4.00 before the beginning of the 2007–2009 financial crisis, by early 2012 has fluctuated around US$3.50. The analysis found that the cost of an equivalent electric-gallon of gasoline would have been not only cheaper to operate during the entire analysis period but also that equivalent electricity prices are more stable and have been declining in terms of equivalent dollars per gallon.

Air Pollution and Greenhouse Gas Emissions

Electric cars, as well as plug-in hybrids operating in all-electric mode, emit no harmful tailpipe pollutants from the onboard source of power, such as particulates (soot), volatile organic compounds, hydrocarbons, carbon monoxide, ozone, lead, and various oxides of nitrogen. The clean air benefit is usually local because, depending on the source of the electricity used to recharge the batteries, air pollutant emissions are shifted to the location of the generation plants. In a similar manner, plug-in electric vehicles operating in all-electric mode do not emit greenhouse gases from the onboard source of power, but from the point of view of a well-to-wheel assessment, the extent of the benefit also depends on the fuel and technology used for electricity generation. This fact has been referred to as the long tailpipe of plug-in electric vehicles. From the perspective of a full life cycle analysis, the electricity used to recharge the batteries must be generated from renewable or clean sources such as wind, solar, hydroelectric, or nuclear power for PEVs to have almost none or zero well-to-wheel emissions. On the other hand, when PEVs are recharged from coal-fired plants, they usually produce slightly more greenhouse gas emissions than internal combustion engine vehicles and higher than hybrid electric vehicles. In the case of plug-in hybrid electric vehicles operating in hybrid mode with assistance of the internal combustion engine, tailpipe and greenhouse emissions are lower in comparison to conventional cars because of their higher fuel economy.

RechargeIT converted Toyota Prius plug-in hybrids at Google's Mountain View campus. The garage has recharging facilities powered by solar panels.

The magnitude of the potential advantage depends on the mix of generation sources and therefore varies by country and by region. For example, France can obtain significant emission benefits from electric and plug-in hybrids because most of its electricity is generated by nuclear power plants; California, where most energy comes from natural gas, hydroelectric and nuclear plants can also secure substantial emission benefits. The U.K. also has a significant potential to benefit from PEVs as natural gas plants dominate the generation mix. On the other hand, emission benefits in Germany, China, India, and the central regions of the United States are limited or non-existent because most electricity is generated from coal. However these countries and regions might still obtain some air quality benefits by reducing local air pollution in urban areas. Cities with chronic air pollution problems, such as Los Angeles, México City, Santiago, Chile, São Paulo, Beijing, Bangkok and Kathmandu may also gain local clean air benefits by shifting the harmful emission to electric generation plants located outside the cities. Nevertheless, the location of the plants is not relevant when considering greenhouse gas emission because their effect is global.

Carbon Footprint during Production

A report published in June 2011, prepared by Ricardo in collaboration with experts from the UK's Low Carbon Vehicle Partnership, found that hybrid electric cars, plug-in hybrids and all-electric cars generate more carbon emissions during their production than current conventional vehicles, but still have a lower overall carbon footprint over the full life cycle. The higher carbon footprint during production of electric drive vehicles is due mainly to the production of batteries. As an example, 43 percent of production emissions for a mid-size electric car are generated from the battery production, while for standard mid-sized gasolineinternal combustion engine vehicle, around 75% of the embedded carbon emissions during production comes from the steel used in the vehicle glider. The following table summarizes key results of this study for four powertrain technologies:

Comparison of full life cycle assessment(well-to-wheels) of carbon emissions and carbon footprint during production for four different powertrain technologies			
Type of vehicle (powertrain)	Estimated emissions in production (tonnes CO2e)	Estimated lifecycle emissions (tonnes CO2e)	Percentage of emissions during production
Standard gasoline vehicle	5.6	24	23%
Hybrid electric vehicle	6.5	21	31%
Plug-in hybrid electric vehicle	6.7	19	35%
Battery electric vehicle	8.8	19	46%
Estimates based upon a 2015 model vehicle assuming 150,000 km (93,000 mi) full life travel using 10% ethanol blend and 500g/kWh grid electricity.			

The Ricardo study also found that the lifecycle carbon emissions for mid-sized gasoline and diesel vehicles are almost identical, and that the greater fuel efficiency of the diesel engine is offset by higher production emissions.

Volkswagen

In 2014 Volkswagen published the results of life-cycle assessment of its electric vehicles certified by TÜV NORD, and independent inspection agency. The study found that CO_2 emissions during the use phase of its all-electric VW e-Golf are 99% lower than those of the Golf 1.2 TSI when powers comes from exclusively hydroelectricity generated in Germany, Austria and Switzerland. Accounting for the full lifecycle, the e-Golf reduces emissions by 61%, offsetting higher production emissions. When the actual EU-27 electricity mix is considered, the e-Golf emissions are still 26% lower than those of the conventional Golf 1.2 TSI. Similar results were found when comparing the e-Golf with the Golf 1.6 TDI. The analysis considered recycling of the three vehicles at the end of their lifetime.

Uniti

The Swedish automotive startup Uniti Sweden AB has been working on an electric city car entirely designed to be sustainable. The CEO Lewis Horne said "*When people sell electric cars, the motivation is sustainability. So, why are they so heavy, and why are they the same cars? Where you just take out petrol and you put in the huge battery and you call it sustainable. But it's not.*". The company has been designing a car that uses sustainable and recycled material such as bio-composite to reduce the footprint of the car during production. In addition, the fully automated factory provided in partnership with Siemens will be able to work with the lights turned off during 22h per day to save energy.

Well-to-wheel GHG Emissions in the U.S.

The following table compares tailpipe and upstream CO_2 emissions estimated by the U.S. Environmental Protection Agency for all series production model year 2014 plug-in electric vehicles available in the U.S. market. Total emissions include the emissions associated with the production and distribution of electricity used to charge the vehicle, and for plug-in hybrid electric vehicles, it also includes emissions associated with tailpipe emissions produced from the internal combustion engine. These figures were published by the EPA in October 2014 in its annual report "*Light-Duty Automotive Technology, Carbon Dioxide Emissions, and Fuel Economy Trends: 1975 Through 2014.*" All emissions are estimated considering average real world city and highway operation based on the EPA 5-cycle label methodology, using a weighted 55% city and 45% highway driving. For the first time, the *2014 Trends* report presents an analysis of the impact of alternative fuel vehicles, with emphasis in plug-in electric vehicles because as their market share is approaching 1%, the EPA concluded that PEVs began to have a measurable impact on the U.S. overall new vehicle fuel economy and CO_2 emissions.

For purposes of an accurate estimation of emissions, the analysis took into consideration the differences in operation between plug-in hybrids. Some, like the Chevrolet Volt, can operate in all-electric mode without using gasoline, and others operate in a blended mode like the Toyota Prius PHV, which uses both energy stored in the battery and energy from the gasoline tank to propel the vehicle, but that can deliver substantial all-electric driving in blended mode. In addition, since the all-electric range of plug-in hybrids depends on the size of the battery pack, the analysis introduced a utility factor as a projection of the share of miles that will be driven using electricity by an average driver, for both, electric only and blended EV modes. Since all-electric cars do not produce tailpipe emissions, the utility factor applies only to plug-in hybrids. The following table shows the overall fuel economy expressed in terms of miles per gallon gasoline equivalent (mpg-e) and the utility factor for the ten MY2014 plug-in hybrids available in the U.S. market, and EPA's best estimate of the CO_2 tailpipe emissions produced by these PHEVs.

In order to account for the upstream CO_2 emissions associated with the production and distribution of electricity, and since electricity production in the United States varies significantly from region to region, the EPA considered three scenarios/ranges with the low end scenario corresponding to the California powerplant emissions factor, the middle of the range represented by the national average powerplant emissions factor, and the upper end of the range corresponding to the powerplant emissions factor for the Rocky Mountains. The EPA estimates that the electricity GHG emission factors for various regions of the country vary from 346 g CO_2/kWh in California to 986 g CO_2/kWh in the Rockies, with a national average of 648 g CO_2/kWh.

Comparison of tailpipe and upstream CO_2 emissions[(1)] estimated by EPA for the MY 2014 plug-in electric vehicles available in the U.S. market						
Vehicle	Overall fuel economy (mpg-e)	Utility factor[(2)] (share EV miles)	Tailpipe CO_2 (g/mi)	Tailpipe + Total Upstream CO2		
				Low (g/mi)	Avg (g/mi)	High (g/mi)
BMW i3	124	1	0	93	175	266
Chevrolet Spark EV	119	1	0	97	181	276
Honda Fit EV	118	1	0	99	185	281
Fiat 500e	116	1	0	101	189	288

Nissan Leaf	114	1	0	104	194	296
Mitsubishi i	112	1	0	104	195	296
Smart electric drive	107	1	0	109	204	311
Ford Focus Electric	105	1	0	111	208	316
Tesla Model S (60 kWh)	95	1	0	122	229	348
Tesla Model S (85 kWh)	89	1	0	131	246	374
BMW i3 REx(3)	88	0.83	40	134	207	288
Mercedes-Benz B-Class ED	84	1	0	138	259	394
Toyota RAV4 EV	76	1	0	153	287	436
BYD e6	63	1	0	187	350	532
Chevrolet Volt	62	0.66	81	180	249	326
Toyota Prius Plug-in Hybrid	58	0.29	133	195	221	249
Honda Accord Plug-in Hybrid	57	0.33	130	196	225	257
Cadillac ELR	54	0.65	91	206	286	377
Ford C-Max Energi	51	0.45	129	219	269	326
Ford Fusion Energi	51	0.45	129	219	269	326
BMW i8	37	0.37	198	303	351	404
Porsche Panamera S E-Hybrid	31	0.39	206	328	389	457
McLaren P1	17	0.43	463	617	650	687
Average MY 2014 gasoline car	24.2	0	367	400	400	400

(1) Based on 45% highway and 55% city driving. (2) The utility factor represents, on average, the percentage of miles that will be driven using electricity (in electric only and blended modes) by an average driver. (3) The EPA classifies the i3 REx as a series plug-in hybrid

The Union of Concerned Scientists (UCS) published a study in 2012 that assessed average greenhouse gas emissions in the U.S. resulting from charging plug-in car batteries from the perspective of the full life-cycle (well-to-wheel analysis) and according to fuel and technology used to generate electric power by region. The study used the model year 2011 Nissan Leaf all-electric car to establish the analysis baseline, and electric-utility emissions are based on EPA's 2009 estimates. The UCS study expressed the results in terms of miles per gallon instead of the conventional unit of grams of greenhouse gases or carbon dioxide equivalent emissions per year in order to make the results more friendly for consumers. The study found that in areas where electricity is generated from natural gas, nuclear, hydroelectric or renewable sources, the potential of plug-in electric cars to reduce greenhouse emissions is significant. On the other hand, in regions where a high proportion of power is generated from coal, hybrid electric cars produce less CO_2 equivalent emissions than plug-in electric cars, and the best fuel efficient gasoline-powered subcompact car produces slightly less emissions than a PEV. In the worst-case scenario, the study estimated that for a region where all energy is generated from coal, a plug-in electric car would emit greenhouse gas emissions equivalent to a gasoline car rated at a combined city/highway driving fuel economy of 30 mpg_{US} (7.8 L/100 km; 36 mpg_{imp}). In contrast, in a region that is completely reliant on natural gas, the PEV would be equivalent to a gasoline-powered car rated at 50 mpg_{US} (4.7 L/100 km; 60 mpg_{imp}).

The study concluded that for 45% of the U.S. population, a plug-in electric car will generate lower CO_2 equivalent emissions than a gasoline-powered car capable of combined 50 mpg_{US} (4.7 L/100 km; 60 mpg_{imp}), such as the Toyota Prius and the Prius c. The UCS also found that for 37% of the population, the electric car emissions will fall in the range of a gasoline-powered car rated at a combined fuel economy of 41 to 50 mpg_{US} (5.7 to 4.7 L/100 km; 49 to 60 mpg_{imp}), such as the Honda Civic Hybrid and the Lexus CT200h. Only 18% of the population lives in areas where the power-supply is more dependent on burning carbon, and the greenhouse gas emissions will be equivalent to a car rated at a combined fuel economy of 31 to 40 mpg_{US} (7.6 to 5.9 L/100 km; 37 to 48 mpg_{imp}), such as the Chevrolet Cruze and Ford Focus. The study found that there are no regions in the U.S. where plug-in electric cars will have higher greenhouse gas emissions than the average new compact gasoline engine automobile, and the area with the dirtiest power supply produces CO_2 emissions equivalent to a gasoline-powered car rated at 33 mpg_{US} (7.1 L/100 km).

In September 2014 the UCS published an updated analysis of its 2012 report. The 2014 analysis found that 60% of Americans, up from 45% in 2009, live in regions where an all-electric car produce fewer CO_2 equivalent emissions per mile than the most efficient hybrid. The UCS study found several reasons for the improvement. First, electric utilities have adopted cleaner sources of electricity to their mix between the two analysis. The 2014 study used electric-utility emissions based on EPA's 2010 estimates, but since coal use nationwide is down by about 5% from 2010 to 2014, actual efficiency in 2014 is better than estimated in the UCS study. Second, electric vehicles have become more efficient, as the average 2013 all-electric vehicle used 0.33 kWh per mile, representing a 5% improvement over 2011 models. Also, some new models are cleaner than the average, such as the BMW i3, which is rated at 0.27 kWh by the EPA. An i3 charged with power from the Midwest grid would be as clean as a gasoline-powered car with about 50 mpg_{US} (4.7 L/100 km), up from 39 mpg_{US} (6.0 L/100 km) for the average electric car in the 2012 study. In states with a cleaner mix generation, the gains were larger. The average all-electric car in California went up to 95 mpg_{US} (2.5 L/100 km) equivalent from 78 mpg_{US} (3.0 L/100 km) in the 2012 study. States with dirtier generation that rely heavily on coal still lag, such as Colorado, where the average BEV only achieves the same emissions as a 34 mpg_{US} (6.9 L/100 km; 41 mpg_{imp}) gasoline-powered car. The author of the 2014 analysis noted that the benefits are not distributed evenly across the U.S. because electric car adoptions is concentrated in the states with cleaner power.

In November 2015 the Union of Concerned Scientists published a new report comparing two battery electric vehicles (BEVs) with similar gasoline vehicles by examining their global warming emissions over their full life-cycle, cradle-to-grave analysis. The two BEVs modeled, midsize and full-size, are based on the two most popular BEV models sold in the United States in 2015, the Nissan Leaf and the Tesla Model S. The study found that all-electric cars representative of those sold today, on average produce less than half the global warming emissions of comparable gasoline-powered vehicles, despite taken into account the higher emissions associated with BEV manufacturing. Considering the regions where the two most popular electric cars are being sold, excess manufacturing emissions are offset within 6 to 16 months of average driving. The study also concluded that driving an average EV results in lower global warming emissions than driving a gasoline car that gets 50 mpg_{US} (4.7 L/100 km) in regions covering two-thirds of the U.S. population, up from 45% in 2009. Based on where EVs are being sold in the United States in 2015, the average EV produces global warming emissions equal to a gasoline vehicle with a 68 mpg_{US} (3.5 L/100 km) fuel economy rating. The authors identified two main reason for the fact that EV-related emissions have become even lower in many parts of the country since the first study was conducted in 2012. Electricity generation has been getting cleaner, as coal-fired generation has declined while lower-carbon alternatives have increased. In addition, electric cars are becoming more efficient. For example, the Nissan Leaf and the Chevrolet Volt, have undergone improvements to increase their efficiencies compared to the original models launched in 2010, and other even more efficient BEV models, such as the most lightweight and efficient BMW i3, have entered the market.

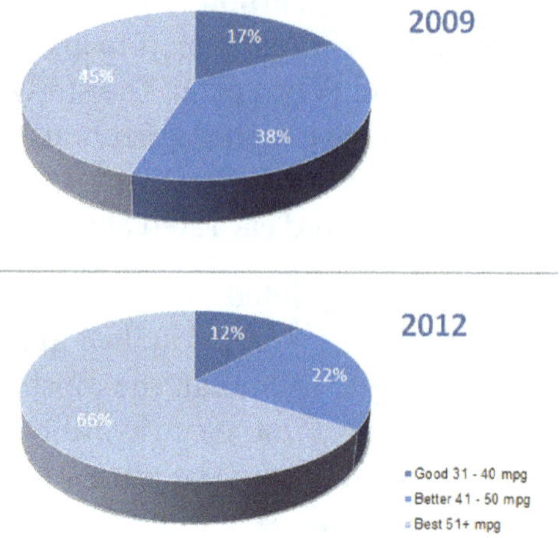

Change from 2009 to 2012 of the percentage of Americans that live in regions where powering an electric vehicle on the regional electricity grid produces lower global warming emissions than a gasoline car expressed in terms of combined city/highway fuel economy rating.

National Bureau of Economic Research

One criticism to the UCS study is that the analysis was made using average emissions rates across regions instead of marginal generation at different times of the day. The former approach does not take into account the generation mix within interconnected electricity markets and shifting load profiles throughout the day. An analysis by three economist affiliated with the National Bureau of Economic Research (NBER), published in November 2014, developed a methodology to estimate marginal emissions of electricity demand that vary by location and time of day across the United States. The marginal analysis, applied to plug-in electric vehicles, found that the emissions of charging PEVs vary by region and hours of the day. In some regions, such as the Western U.S. and Texas, CO_2 emissions per mile from driving PEVs are less than those from driving a hybrid car. However, in other regions, such as the Upper Midwest, charging during the recommended hours of midnight to 4 a.m. implies that PEVs generate more emissions per mile than the average car currently on the road. The results show a fundamental tension between electricity load management and environmental goals as the hours when electricity is the least expensive to produce tend to be the hours with the greatest emissions. This occurs because coal-fired units, which have higher emission rates, are most commonly used to meet base-level and off-peak electricity demand; while natural gas units, which have relatively low emissions rates, are often brought online to meet peak demand.

Well-to-wheel GHG Emissions in Several Countries

A study published in the UK in April 2013 assessed the carbon footprint of plug-in electric vehicles in 20 countries. As a baseline the analysis established that manufacturing emissions account for 70 g CO_2/km for an electric car and 40 g CO_2/km for a petrol car. The study found that in countries with coal-intensive generation, PEVs are no different from conventional petrol-powered vehicles. Among these countries are China, Indonesia, Australia, South Africa and India. A pure electric car in India generates emissions comparable to a 20 mpg_{US} (12 L/100 km; 24 mpg_{imp}) petrol car.

The country ranking was led by Paraguay, where all electricity is produced from hydropower, and Iceland, where electricity production relies on renewable power, mainly hydro and geothermal power. Resulting carbon emissions from an electric car in both countries are 70 g CO_2/km, which is equivalent to a 220 mpg_{US} (1.1 L/100 km; 260 mpg_{imp}) petrol car, and correspond to manufacturing emissions. Next in the ranking are other countries with low carbon electricity generation, including Sweden (mostly hydro and nuclear power), Brazil (mainly hydropower) and France (predominantly nuclear power). Countries ranking in the middle include Japan, Germany, the UK and the United States.

The following table shows the emissions intensity estimated in the study for those countries where electric vehicle are available, and the corresponding emissions equivalent in miles per US gallon of a petrol-powered car:

Country comparison of full life cycle assessment of greenhouse gas emissions resulting from charging plug-in electric cars and emissions equivalent in terms of miles per US gallon of a petrol-powered car				
Country	PEV well-to-wheels carbon dioxide equivalent emissions per electric car expressed in (CO_2e/km)	Power source	PEV well-to-wheels emissions equivalent in terms of mpg US of petrol-powered car	Equivalent petrol car
Sweden	81	Low carbon	159 mpg_{US} (1.48 L/100 km)	Hybrid multiples
France	93		123 mpg_{US} (1.91 L/100 km)	
Canada	115	Fossil light	87 mpg_{US} (2.7 L/100 km)	Beyond hybrid
Spain	146		61 mpg_{US} (3.9 L/100 km)	
Japan	175	Broad mix	48 mpg_{US} (4.9 L/100 km)	New hybrid
Germany	179		47 mpg_{US} (5.0 L/100 km)	
United Kingdom	189		44 mpg_{US} (5.3 L/100 km)	
United States	202	Fossil heavy	40 mpg_{US} (5.9 L/100 km)	Efficient petrol
Mexico	203		40 mpg_{US} (5.9 L/100 km)	
China	258	Coal-based	30 mpg_{US} (7.8 L/100 km)	Average petrol
Australia	292		26 mpg_{US} (9.0 L/100 km)	
India	370		20 mpg_{US} (12 L/100 km)	
Electric car manufacturing emissions account for 70 g CO_2/km				

Less Dependence on Imported Oil

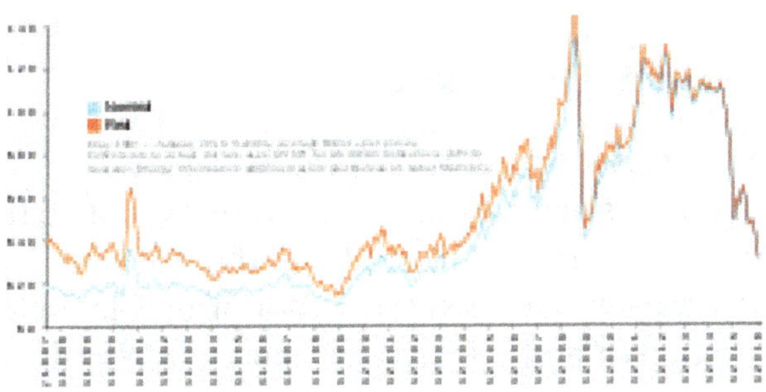

Evolution of oil prices since 1987 (average Brent spot prices – adjusted for U.S. inflation).

For many net oil importing countries the 2000s energy crisis brought back concerns first raised during the 1973 oil crisis. For the United States, the other developed countries and emerging countries their dependence on foreign oil has revived concerns about their vulnerability to price shocks and supply disruption. Also, there have been concerns about the uncertainty surrounding peak oil production and the higher cost of extracting unconventional oil. A third issue that has been raised is the threat to national security because most proven oil reserves are concentrated in relatively few geographic locations, including some countries with strong resource nationalism, unstable governments or hostile to U.S. interests. In addition, for many developing countries, and particularly for the poorest African countries, high oil prices have an adverse impact on the government budget and deteriorate their terms of trade thus jeopardizing their balance of payments, all leading to lower economic growth.

Through the gradual replacement of internal combustion engine vehicles for electric cars and plug-in hybrids, electric drive vehicles can contribute significantly to lessen the dependence of the transport sector on imported oil as well as contributing to the development of a more resilient energy supply.

Vehicle-to-grid

Plug-in electric vehicles offer users the opportunity to sell electricity stored in their batteries back to the power grid, thereby helping utilities to operate more efficiently in the management of their demand peaks. A vehicle-to-grid (V2G) system would take advantage of the fact that most vehicles are parked an average of 95 percent of the time. During such idle times the electricity stored in the batteries could be transferred from the PEV to the power lines and back to the grid. In the U.S. this transfer back to the grid have an estimated value to the utilities of up to $4,000 per year per car. In a V2G system it would also be expected that battery electric (BEVs) and plug-in hybrids (PHEVs) would have the capability to communicate automatically with the power grid to sell demand response services by either delivering electricity into the grid or by throttling their charging rate.

Disadvantages

Tesla Model S electric car (left) and Fisker Karma plug-in hybrid (right) at the parking spots reserved for green cars at San Francisco International Airport.

Cost of Batteries and Cost of Ownership

As of 2015, plug-in electric vehicles are significantly more expensive as compared to conventional internal combustion engine vehicles and hybrid electric vehicles due to the additional cost of their lithium-ion battery pack. According to a 2010 study by the National Research Council, the cost of a lithium-ion battery pack was about US$1,700/kWh of usable energy, and considering that a PHEV-10 requires about 2.0 kWh and a PHEV-40 about 8 kWh, the manufacturer cost of the battery pack for a PHEV-10 is around US$3,000 and it goes up to US$14,000 for a PHEV-40. As of June 2012, and based on the three battery size options offered for the Tesla Model S, the New York Times estimated the cost of automotive battery packs between US$400 to US$500 per kilowatt-hour. A 2013 study by the American Council for an Energy-Efficient Economy reported that battery costs came down from US$1,300 per kWh in 2007 to US$500 per kWh in 2012. The U.S. Department of Energy has set cost targets for its sponsored battery research of US$300 per kWh in 2015 and US$125 per kWh by 2022. Cost reductions through advances in battery technology and higher production volumes will allow plug-in electric vehicles to be more competitive with conventional internal combustion engine vehicles.

According to a study published in February 2016 by Bloomberg New Energy Finance (BNEF), battery prices fell 65% since 2010, and 35% just in 2015, reaching US$350 per kWh. The study concludes that battery costs are on a trajectory to make electric vehicles without government subsidies as affordable as internal combustion engine cars in most countries by 2022. BNEF projects that by 2040, long-range electric cars will cost less than US$22,000 expressed in 2016 dollars. BNEF expects electric car battery costs to be well below US$120 per kWh by 2030, and to fall further thereafter as new chemistries become available.

Cost of Ownership

A study published in 2011 by the Belfer Center, Harvard University, found that the gasoline costs savings of plug-in electric cars do not offset their higher purchase prices when comparing their lifetime net present value of purchase and operating costs for the U.S. market at 2010 prices, and assuming no government subidies. According to the study estimates, a PHEV-40 is US$5,377 more expensive than a conventional internal combustion engine, while a battery electric vehicles is US$4,819 more expensive. These findings assumed a battery cost of US$600 per kWh, which means that the Chevrolet Volt battery pack cost around US$10,000 and the Nissan Leaf pack costs US$14,400. The study also assumed a gasoline price of US$3.75 per gallon (as of mid June 2011), that vehicles are driven 12,000 miles (19,000 km) per year, an average price of electricity of US$0.12 per kWh, that the plug-in hybrid is driven in all-electric mode 85% of the time, and that the owner of PEVs pay US$1,500 to install a Level II 220/240 volt charger at home.

The study also include hybrid electric vehicles in the comparison, and analyzed several scenarios to determine how the comparative net savings will change over the next 10 to 20 years, assuming that battery costs will decrease while gasoline prices increase, and also assuming higher fuel efficiency of conventional cars, among other scenarios. Under the future scenarios considered, the study found that BEVs will be significantly less expensive than conventional cars (US$1,155 to US$7,181 cheaper), while PHEVs, will be more expensive than BEVs in almost all comparison scenarios, and only less expensive than conventional cars in a scenario with very low battery costs and high gasoline prices. The reason for the different savings among PEVs is because BEVs are simpler to build and do not use liquid fuel, while PHEVs have more complicated powertrains and still have gasoline-powered engines.

According to a study by the Electric Power Research Institute published in June 2013, the total cost of ownership of the 2013 Nissan Leaf SV is substantially lower than that of comparable conventional and hybrid vehicles. For comparison, the study constructed average hybrid and conventional vehicles and assumed an average US distance per trip distribution. The study took into account the manufacturer's suggested retail price, taxes, credits, destination charge, electric charging station, fuel cost, maintenance cost, and additional cost due to the use of a gasoline vehicle for trips beyond the range of the Leaf.

Electric Power Research Institute comparison of the Leaf versus average conventional and hybrid cars.			
Vehicle	Operating mode (powertrain)	Total ownership cost	
		US Average	California
Nissan Leaf SV	All-electric	\$37,288	\$35,596
Chevrolet Volt	Plug-in hybrid	\$44,176	\$40,800
Average Conventional	Gasoline	\$44,949	\$46,561
Average Hybrid	Gasoline-electric hybrid	\$44,325	\$45,416
Costs are based on a gasoline price of \$3.64 per gallon, an electricity rate of \$0.12/kWh, and a vehicle lifetime of 150,000 miles. The average conventional car was constructed by averaging of Honda Civic EX, Chevrolet Cruze LTZ, Ford Focus Titanium, and Volkswagen Passat. The average hybrid car was constructed from Ford Fusion Hybrid, Honda Civic Hybrid, Toyota Camry Hybrid XLE, and Toyota Prius trim IV.			

Availability of Recharging Infrastructure

Converted Toyota Prius plug-in hybrids recharging at San Francisco City Hall charging station.

Despite the widespread assumption that plug-in recharging will take place overnight at home, residents of cities, apartments, dormitories, and townhouses do not have garages or driveways with available power outlets, and they might be less likely to buy plug-in electric vehicles unless recharging infrastructure is developed. Electrical outlets or charging stations near their places of residence, in commercial or public parking lots, streets and workplaces are required for these potential users to gain the full advantage of PHEVs, and in the case of EVs, to avoid the fear of the batteries running out energy before reaching their destination, commonly called range anxiety. Even house dwellers might need to charge at the office or to take advantage of opportunity charging at shopping centers. However, this infrastructure is not in place and it will require investments by both the private and public sectors.

Several cities in California and Oregon, and particularly San Francisco and other cities in the San Francisco Bay Area and Silicon Valley, already have deployed public charging stations and have expansion plans to attend both plug-ins and all-electric cars. Some local private firms such as Google and Adobe Systems have also deployed charging infrastructure. In Google's case, its Mountain

View campus has 100 available charging stations for its share-use fleet of converted plug-ins available to its employees. Solar panels are used to generate the electricity, and this pilot program is being monitored on a daily basis and performance results are published on the RechargeIT website. As of December 2013, Estonia is the first and only country that had deployed an EV charging network with nationwide coverage, with 165 fast chargers available along highways at a minimum distance of between 40 to 60 km (25 to 37 mi), and a higher density in urban areas.

The importance to build the infrastructure necessary to support electric vehicles is illustrated by the decision of Car2Go in San Diego, California, that due to insufficient charging infrastructure decided to replace all of its all-electric car fleet with gasoline-powered cars starting on 1 May 2016. When the carsharing service started in 2011, Car2Go expected 1,000 charging stations to be deployed around the city, but only 400 were in place by early 2016. As a result, an average of 20% of the carsharing fleet is unavailable at any given time because the cars are either being charged or because they don't have enough electricity in them to be driven. Also, many of the company's 40,000 San Diego members say they often worry their Car2Go will run out of charge before they finish their trip.

Battery Swapping

Better Place's battery switching station in Israel

A different approach to resolve the problems of range anxiety and lack of recharging infrastructure for electric vehicles was developed by Better Place. Its business model considers that electric cars are built and sold separately from the battery pack. As customers are not allowed to purchase battery packs, they must lease them from Better Place which will deploy a network of battery swapping stations thus expanding EVs range and allowing long distance trips. Subscribed users pay a per-distance fee to cover battery pack leasing, charging and swap infrastructure, the cost of sustainable electricity, and other costs. Better Place signed agreement for deployment in Australia, Denmark, Israel, Canada, California, and Hawaii. The Renault Fluence Z.E. was the electric car built with switchable battery technology sold for the Better Place network. The robotic battery-switching operation was completed in about five minutes.

After implementing the first modern commercial deployment of the battery swapping model in Israel and Denmark, Better Place filed for bankruptcy in Israel in May 2013. The company's financial difficulties were caused by the high investment required to develop the charging and swapping infrastructure, about US$850 million in private capital, and a market penetration significantly lower than originally predicted by Shai Agassi. Less than 1,000 Fluence Z.E. cars were deployed in Israel and around 400 units in Denmark.

Tesla Motors designed its Model S to allow fast battery swapping. In June 2013, Tesla announced their goal to deploy a battery swapping station in each of its supercharging stations. At a demonstration event Tesla showed that a battery swap operation with the Model S takes just over 90 seconds, about half the time it takes to refill a gasoline-powered car used for comparison purposes during the event. The first stations are planned to be deployed along Interstate 5 in California where, according to Tesla, a large number of Model S sedans make the San Francisco-Los Angeles trip regularly. These will be followed by the Washington, DC to Boston corridor.

Other Charging Solutions

Roof-mounted solar panels of the REVA NXR concept car.

Nissan Leaf SV roof-mounted solar panel.

The REVA NXR exhibited in the 2009 Frankfurt Motor Show and the Nissan Leaf SV trim both have roof-mounted solar panels. These solar panels are designed to trickle charge the batteries when the car is moving or parked. Another proposed technology is REVive, by Reva. When the REVA NXR's batteries are running low or are fully depleted, the driver is able to send an SMS to REVive and unlock a hidden reserve in the battery pack. Reva has not provided details on how the system will work. The Fisker Karma uses solar panel in the roof to recharge the 12-volt lead-acid accessory battery. The Nissan Leaf SL trim also has a small solar panel at the rear of the roof/spoiler that can trickle charge the auxiliary 12-volt lead-acid battery.

Potential Overload of the Electrical Grid

The existing electrical grid, and local transformers in particular, may not have enough capacity to handle the additional power load that might be required in certain areas with high plug-in electric car concentrations. As recharging a single electric-drive car could consume three times as much electricity as a typical home, overloading problems may arise when several vehicles in the same neighborhood recharge at the same time, or during the normal summer peak loads. To avoid such problems, utility executives recommend owners to charge their vehicles overnight when the grid load is lower or to use smarter electric meters that help control demand. When market penetration of plug-in electric vehicles begins to reach significant levels, utilities will have to invest in improvements for local electrical grids in order to handle the additional loads related to recharging to avoid blackouts due to grid overload. Also, some experts have suggested that by implementing variable time-of-day rates, utilities can provide an incentive for plug-in owners to recharge mostly overnight, when rates are lower.

General Motors is sponsoring the Pecan Street demonstration project in Austin, Texas. The project objective is to learn the charging patterns of plug-in electric car owners, and to study how a residential fleet of electric vehicles might strain the electric grid if all owners try to charge them at the same, which is what the preliminary monitoring found when the plug-in cars return home in the evening. The Mueller neighborhood is the test ground, and as of June 2013, the community has nearly 60 Chevrolet Volt owners alone. This cluster of Volts was achieved thanks to GM's commitment to match the federal government's $7,500 rebate incentive, which effectively halves the purchase price of the plug-hybrid electric cars.

Risks Associated with Noise Reduction

The 2011 Nissan Leaf had a switch to manually turn off its electric warning sound system.

Electric cars and plug-in hybrids when operating in all-electric mode at low speeds produce less roadway noise as compared to vehicles propelled by an internal combustion engine, thereby reducing harmful noise health effects. However, blind people or the visually impaired consider the noise of combustion engines a helpful aid while crossing streets, hence plug-in electric cars and conventional hybrids could pose an unexpected hazard when operating at low speeds. Several tests conducted in the U.S. have shown that this is a valid concern, as vehicles operating in electric mode can be particularly hard to hear below 20 mph (30 km/h) for all types of road users and not only the visually impaired. At higher speeds the sound created by tire friction and the air displaced by the vehicle start to make sufficient audible noise.

Some carmakers announced they have decided to address this safety issue, and as a result, the Nissan Leaf electric car and Chevrolet Volt plug-in hybrid, both launched in December 2010, as well as the Fisker Karma plug-in hybrid launched in 2011 launched in 2012, include electric warning sounds to alert pedestrians, the blind and others to their presence. As of January 2014, most of the hybrids and plug-in electric and hybrids available in the United States, Japan and Europe make warning noises using a speaker system.

The Japanese Ministry of Land, Infrastructure, Transport and Tourism issued guidelines for hybrid and other near-silent vehicles in January 2010. In the United States the Pedestrian Safety Enhancement Act of 2010 was approved by the U.S. Senate and the House of Representatives in December 2010. After several delays, the National Highway Traffic Safety Administration (NHTSA) issued its ruling in February 2018. It requires hybrids and electric vehicles traveling at less than 18.6 mph (30 km/h) to emit warning sounds that pedestrians must be able to hear over background noises. The regulation requires full compliance in September 2020, but 50% of "quiet" vehicles must have the warning sounds by September 2019.

In April 2014 the European Parliament approved legislation that requires the mandatory use of the AVAS for all new electric and hybrid electric vehicles and car manufacturers have to comply within 5 years.

Risks of Battery Fire

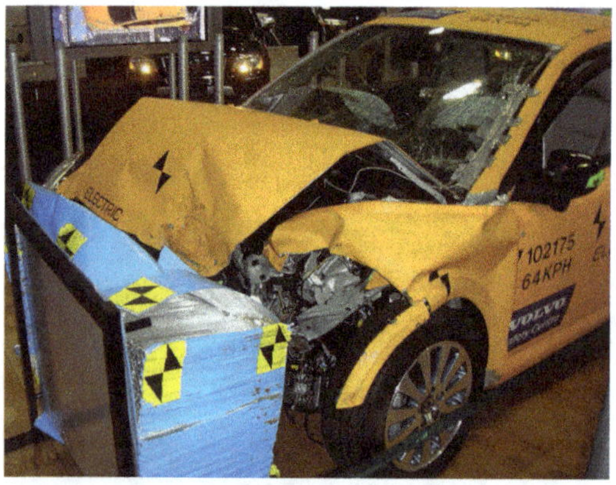

Frontal crash test of a Volvo C30 DRIVe Electric to assess the safety of the battery pack.

Lithium-ion batteries may suffer thermal runaway and cell rupture if overheated or overcharged, and in extreme cases this can lead to combustion. To reduce these risks, lithium-ion battery packs contain fail-safe circuitry that shuts down the battery when its voltage is outside the safe range. When handled improperly, or if manufactured defectively, some rechargeable batteries can experience thermal runaway resulting in overheating. Especially prone to thermal runaway are lithium-ion batteries. Reports of exploding cellphones have been reported in newspapers. In 2006, batteries from Apple, HP, Toshiba, Lenovo, Dell and other notebook manufacturers were recalled because of fire and explosions. Also, during the Boeing 787 Dreamliner's first year of service, at least four aircraft suffered from electrical system problems stemming from its lithium-ion batteries, resulting in the whole Dreamliner fleet being voluntarily grounded in January 2013.

Several plug-in electric vehicle fire incidents have taken place since the introduction of mass-production plug-in electric vehicles in 2008. Most of them have been thermal runaway incidents related to the lithium-ion batteries and have involved the Zotye M300 EV, Chevrolet Volt, Fisker Karma, BYD e6, Dodge Ram 1500 Plug-in Hybrid, Toyota Prius Plug-in Hybrid, Mitsubishi i-MiEV and Outlander P-HEV. As of November 2013, four fires after a crash have been reported associated with the batteries of all-electric cars involving a BYD e6 and three Tesla Model S cars.

The first modern crash-related fire was reported in China in May 2012, after a high-speed car crashed into a BYD e6 taxi in Shenzhen. The second reported incident occurred in the United States on October 1, 2013, when a Tesla Model S caught fire after the electric car hit metal debris on a highway in Kent, Washington state, and the debris punctured one of 16 modules within the battery pack. A second reported fire occurred on October 18, 2013, in Merida, Mexico. In this case the vehicle was being driven at high speed through a roundabout and crashed through a wall and into a tree. On November 6, 2013, a Tesla Model S being driven on Interstate 24 near Murfreesboro, Tennessee caught fire after it struck a tow hitch on the roadway, causing damage beneath the vehicle.

The U.S. National Highway Traffic Safety Administration (NHTSA) is conducting a study due in 2014 to establish whether lithium-ion batteries in plug-electric vehicles pose a potential fire hazard. The research is looking at whether the high-voltage batteries can cause fires when they are being charged and when the vehicles are involved in an accident. Both General Motors and Nissan have published a guide for firefighters and first responders to properly handle a crashed plug-in electric-drive vehicle and safely disable its battery and other high voltage systems.

Rare-earth Metals Availability and Supply Security

Common technology for plug-ins and electric cars is based on the lithium-ion battery and an electric motor which uses rare-earth elements. The demand for lithium, heavy metals, and other specific elements (such as neodymium, boron and cobalt) required for the batteries and powertrain is expected to grow significantly due to the future sales increase of plug-in electric vehicles in the mid and long term. As of 2011, the Toyota Prius battery contains more than 20 lb (9.1 kg) of the rare-earth element lanthanum, and its motor magnets use neodymium and dysprosium. While only 0.25 oz (7 g) of lithium carbonate equivalent (LCE) are required in a smartphone and 1.1 oz (30 g) in a tablet computer, electric vehicles and stationary energy storage systems for homes, businesses or industry use much more lithium in their batteries. As of 2016 a hybrid electric passenger car might use 11 lb (5 kg) of LCE, while one of Tesla's high performance electric cars could use as much as 180 lb (80 kg).

Some of the largest world reserves of lithium and other rare metals are located in countries with strong resource nationalism, unstable governments or hostility to U.S. interests, raising concerns about the risk of replacing dependence on foreign oil with a new dependence on hostile countries to supply strategic materials.

Lithium

The main deposits of lithium are found in China and throughout the Andes mountain chain in South America. In 2008 Chile was the leading lithium metal producer with almost 30%, followed by China, Argentina, and Australia. In the United States lithium is recovered from brine pools in Nevada.

Nearly half the world's known reserves are located in Bolivia, and according to the US Geological

Survey, Bolivia's Salar de Uyuni desert has 5.4 million tons of lithium. Other important reserves are located in Chile, China, and Brazil. Since 2006 the Bolivian government have nationalized oil and gas projects and is keeping a tight control over mining its lithium reserves. Already the Japanese and South Korean governments, as well as companies from these two countries and France, have offered technical assistance to develop Bolivia's lithium reserves and are seeking to gain access to the lithium resources through a mining and industrialization model suitable to Bolivian interests.

The Salar de Uyuni in Bolivia is one of the largest known lithium reserves in the world.

According to a 2011 study conducted at Lawrence Berkeley National Laboratory and the University of California Berkeley, the currently estimated reserve base of lithium should not be a limiting factor for large-scale battery production for electric vehicles, as the study estimated that on the order of 1 billion 40 kWh Li-based batteries (about 10 kg of lithium per car) could be built with current reserves, as estimated by the U.S. Geological Survey. Another 2011 study by researchers from the University of Michigan and Ford Motor Company found that there are sufficient lithium resources to support global demand until 2100, including the lithium required for the potential widespread use of hybrid electric, plug-in hybrid electric and battery electric vehicles. The study estimated global lithium reserves at 39 million tons, and total demand for lithium during the 90-year period analyzed at 12–20 million tons, depending on the scenarios regarding economic growth and recycling rates.

A 2016 study by Bloomberg New Energy Finance (BNEF) found that availability of lithium and other finite materials used in the battery packs will not be a limiting factor for the adoption of electric vehicles. BNEF estimated that battery packs will require less than 1% of the known reserves of lithium, nickel, manganese, and copper through 2030, and 4% of the world's cobalt. After 2030, the study states that new battery chemistries will probably shift to other source materials, making packs lighter, smaller, and cheaper.

Rare-earth Elements

China has 48% of the world's reserves of rare-earth elements, the United States has 13%, and Russia, Australia, and Canada have significant deposits. Until the 1980s, the U.S. led the world in rare-earth production, but since the mid-1990s China has controlled the world market for these elements. The mines in Bayan Obo near Baotou, Inner Mongolia, are currently the largest source of rare-earth metals and are 80% of China's production. In 2010 China accounted for 97% of the global production of 17 rare-earth elements. Since 2006 the Chinese government has been imposing export quotas reducing supply at a rate of 5% to 10% a year.

Prices of several rare-earth elements increased sharply by mid-2010 as China imposed a 40% export reduction, citing environmental concerns as the reason for the export restrictions. These quotas have been interpreted as an attempt to control the supply of rare earths. However, the high prices have provided an incentive to begin or reactivate several rare-earth mining projects around the world, including the United States, Australia, Vietnam, and Kazakhstan.

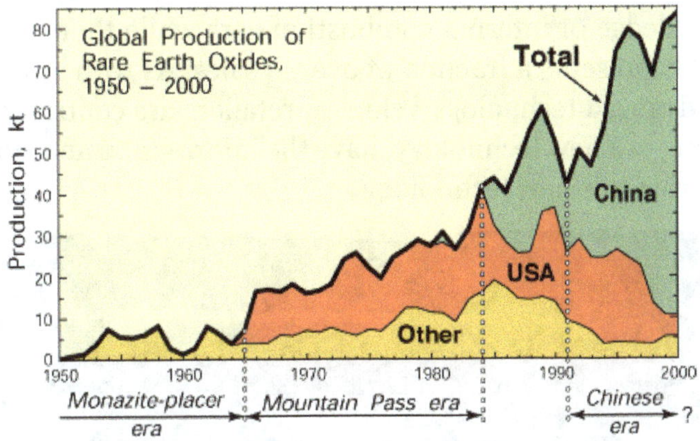

Evolution of global rare-earth oxides production by country (1950–2000)

In September 2010, China temporarily blocked all exports of rare earths to Japan in the midst of a diplomatic dispute between the two countries. These minerals are used in hybrid cars and other products such wind turbines and guided missiles, thereby augmenting the worries about the dependence on Chinese rare-earth elements and the need for geographic diversity of supply. A December 2010 report published by the US DoE found that the American economy vulnerable to rare-earth shortages and estimates that it could take 15 years to overcome dependence on Chinese supplies. China raised export taxes for some rare earths from 15 to 25%, and also extended taxes to exports of some rare-earth alloys that were not taxed before. The Chinese government also announced further reductions on its export quotas for the first months of 2011, which represent a 35% reduction in tonnage as compared to exports during the first half of 2010.

On September 29, 2010, the U.S. House of Representatives approved the Rare Earths and Critical Materials Revitalization Act of 2010. The approved legislation is aimed at restoring the U.S. as a leading producer of rare-earth elements, and would support activities in the U.S. Department of Energy (US DoE) to discover and develop rare-earth sites inside of the U.S. in an effort to reduce the auto industry's near-complete dependence on China for the minerals. A similar bill, the Rare Earths Supply Technology and Resources Transformation Act of 2010, is being discussed in the U.S. Senate.

In order to avoid its dependence on rare-earth minerals, Toyota Motor Corporation announced in January 2011 that it is developing an alternative motor for future hybrid and electric cars that does not need rare-earth materials. Toyota engineers in Japan and the U.S. are developing an induction motor that is lighter and more efficient than the magnet-type motor used in the Prius, which uses two rare earths in its motor magnets. Other popular hybrids and plug-in electric cars in the market that use these rare-earth elements are the Nissan Leaf, the Chevrolet Volt and Honda Insight. For its second generation RAV4 EV due in 2012, Toyota is using an induction motor supplied by Tesla Motors that does not require rare-earth materials. The Tesla Roadster and the Tesla Model S use a similar motor.

Car Dealers Reluctance to Sell

With the exception of Tesla Motors, almost all new cars in the United States are sold through dealerships, so they play a crucial role in the sales of electric vehicles, and negative attitudes can hinder early adoption of plug-in electric vehicles. Dealers decide which cars they want to stock, and a salesperson can have a big impact on how someone feels about a prospective purchase. Sales people have ample knowledge of internal combustion cars while they do not have time to learn about a technology that represents a fraction of overall sales. As with any new technology, and in the particular case of advanced technology vehicles, retailers are central to ensuring that buyers, especially those switching to a new technology, have the information and support they need to gain the full benefits of adopting this new technology.

Car dealerships play a crucial role in the sales of plug-in electric vehicles. Shown a dealership exhibiting first generation Chevrolet Volts.

There are several reasons for the reluctance of some dealers to sell plug-in electric vehicles. PEVs do not offer car dealers the same profits as gasoline-powered car. Plug-in electric vehicles take more time to sell because of the explaining required, which hurts overall sales and sales people commissions. Electric vehicles also may require less maintenance, resulting in loss of service revenue, and thus undermining the biggest source of dealer profits, their service departments. According to the National Automobile Dealers Association (NADS), dealers on average make three times as much profit from service as they do from new car sales. However, a NADS spokesman said there was not sufficient data to prove that electric cars would require less maintenance. According to the New York Times, BMW and Nissan are among the companies whose dealers tend to be more enthusiastic and informed, but only about 10% of dealers are knowledgeable on the new technology.

A study conducted at the Institute of Transportation Studies (ITS), at the University of California, Davis (UC Davis) published in 2014 found that many car dealers are less than enthusiastic about plug-in vehicles. ITS conducted 43 interviews with six automakers and 20 new car dealers selling plug-in vehicles in California's major metro markets. The study also analyzed national and state-level J.D. Power 2013 Sales Satisfaction Index (SSI) study data on customer satisfaction with new car dealerships and Tesla retail stores. The researchers found that buyers of plug-in electric vehicles were significantly less satisfied and rated the dealer purchase experience much lower than buyers of non-premium conventional cars, while Tesla Motors earned industry-high scores.

According to the findings, plug-in buyers expect more from dealers than conventional buyers, including product knowledge and support that extends beyond traditional offerings.

In 2014 Consumer Reports published results from a survey conducted with 19 secret shoppers that went to 85 dealerships in four states, making anonymous visits between December 2013 and March 2014. The secret shoppers asked a number of specific questions about cars to test the salespeople's knowledge about electric cars. The consumer magazine decided to conduct the survey after several consumers who wanted to buy a plug-in car reported to the organization that some dealerships were steering them toward gasoline-powered models. The survey found that not all sales people seemed enthusiastic about making PEV sales; a few outright discouraged it, and even one dealer was reluctant to even show a plug-in model despite having one in stock. And many sales people seemed not to have a good understanding of electric-car tax breaks and other incentives or of charging needs and costs. Consumer Reports also found that when it came to answering basic questions, sales people at Chevrolet, Ford, and Nissan dealerships tended to be better informed than those at Honda and Toyota. The survey found that most of the Toyota dealerships visited recommended against buying a Prius Plug-in and suggested buying a standard Prius hybrid instead. Overall, the secret shoppers reported that only 13 dealers "discouraged sale of EV," with seven of them being in New York. However, at 35 of the 85 dealerships visited, the secret shoppers said sales people recommended buying a gasoline-powered car instead.

The ITS-Davis study also found that a small but influential minority of dealers have introduced new approaches to better meet the needs of plug-in customers. Examples include marketing carpool lane stickers, enrolling buyers in charging networks, and preparing incentive paperwork for customers. Some dealers assign seasoned sales people as plug-in experts, many of whom drive plug-ins themselves to learn and be familiar with the technology and relate the car's benefits to potential buyers. The study concluded also that carmakers could do much more to support dealers selling PEVs.

Production Plug-in Electric Vehicles Available

The General Motors EV1 was the first mass-produced and purpose-designed all-electric car of the modern era from a major automaker.

During the 1990s several highway-capable plug-in electric cars were produced in limited quantities, all were battery electric vehicles. PSA Peugeot Citroën launched several electric

"Électrique" versions of its models starting in 1991, notably the Citroën C15, C25 and Berlingo and Peugeot J5 and Partner panel vans and the Citroën AX and Saxo and Peugeot 106 superminis. The Saxo was the best-selling and most produced ever electric car at the time. Other models were available through leasing mainly in California. Popular models included the General Motors EV1 and the Toyota RAV4 EV. Some of the latter were sold to the public and are in use still today. In the late 2000s began a new wave of mass production plug-in electric cars, motorcycles and light trucks. However, as of 2011, most electric vehicles in the world roads were low-speed, low-range neighborhood electric vehicles (NEVs) or electric quadricycles. Pike Research estimated there were almost 479,000 NEVs on the world roads in 2011. As of October 2015, the GEM neighborhood electric vehicle is the market leader in North America, with global sales of more than 50,000 units since 1998. Sales of low-speed electric vehicles experienced considerable growth in China between 2012 and 2016. As of December 2016, the Chinese stock of NEVs was estimated to be between 3 million and 4 million units, with most powered by lead-acid batteries.

Two plug-in electric cars, a Honda Fit EV all-electric car (left) and a Ford C-Max Energi plug-in hybrid (right), charging from an on-street public charging station in San Francisco.

As of December 2016, there were over 60 models of highway-capable plug-in electric passenger cars and light-utility vans available in the world. As of August 2015, there were 45 different plug-in electric passenger car models offered in Europe, 20 available in North America, 19 in China, 14 in Japan, and 7 in Australia. There are also available several commercial models of plug-in motorcycles, all-electric buses, and heavy-duty trucks.

Cumulative global sales of highway legal plug-in electric passenger cars and light utility vehicles passed the one million unit milestone in September 2015, the two million mark in December 2016, and the three million milestone was achieved in November 2017. The Renault-Nissan-Mitsubishi Alliance is the world's leading electric vehicle manufacturer. The Alliance reached sales of 500,000 electric vehicles delivered globally in October 2017, including the models manufactured by Mitsubishi Motors, now part of the Alliance.

BYD Auto is the world's second largest plug-in electric car manufacturer with more than 300,000 units delivered in China through December 2017. Its Qin plug-in hybrid is the company's top selling model with over 89,000 units sold in China through December 2017, making it the all-time best-selling plug-in electric car in the country. As of September 2017,

Tesla ranked as the world's third largest plug-in electric vehicle manufacturer with global sales of over 250,000 units in September 2017. Its Model S was the world's top selling plug-in car for two years running, 2015 and 2016, and Model S global sales passed the 200,000 unit mark in the fourth quarter of 2017. As of April 2018, the BMW Group had sold over 250,000 plug-in cars, accounting for global sales its BMW i cars, BMW iPerformance plug-in hybrid models, and MINI brand plug-ins.

The Tesla Model S was the world's top selling plug-in car for two years running, 2015 and 2016.

BYD Auto ended 2015 as the world's best selling manufacturer of highway legal light-duty plug-in electric vehicles, with 61,722 units sold, mostly plug-in hybrids, followed by Tesla, with 50,580 units sold in 2015. BYD was the world's top selling plug-in car manufacturer for a second year running, with 101,183 units sold in 2016, one more time followed again by Tesla with 76,243 units delivered. In 2017 BYD ranked for the third year in a row as the global top plug-in car manufacturer with 113,669 units delivered. The Renault-Nissan-Mitsubishi Alliance was the top selling automotive group with 119,995 units sold worldwide in 2017.

Sales and Main Markets

Detroit Electric car charging in 1919. During the Golden Age of the electric car at the beginning of the 20th century, the EV stock peaked at about 30,000 vehicles.

The global stock of plug-in electric vehicles between 2005 and 2009 consisted exclusively of all-electric cars, totaling about 1,700 units in 2005, and almost 6,000 in 2009. The plug-in stock rose to about 12,500 units in 2010, of which, only 350 vehicles were plug-in hybrids. By comparison, during the Golden Age of the electric car at the beginning of the 20th century, the EV stock peaked at approximately 30,000 vehicles. After the introduction of the Nissan Leaf and the Chevrolet Volt in late December 2010, the first mass-production plug-in cars by major carmakers, plug-in car sales grew to about 50,000 units in 2011, jumped to 125,000 in 2012, and rose to almost 213,000 plug-in electric cars and utility vans in 2013. Sales totaled over 315,000 units in 2014, up 48% from 2013, By mid-September 2015, the global stock of highway legal plug-in electric passenger cars and utility vans reached the one million sales milestone. Sales of plug-in electric vehicles achieved the one million milestone almost twice as fast as hybrid electric vehicles (HEV). While it took four years and 10 months to reach one-million PEV sales, it took more than around nine years and a few months for HEVs to reach its first million sales. A 2016 analysis by the Consumer Federation of America (CFA) found that five years after its introduction, sales of plug-in electric cars in the American market outsold conventional hybrids during the same period.

The global ratio between all-electrics (BEVs) and plug-in hybrids (PHEVs) was 60:40 between 2014 and the first half of 2016, mainly due to the large all-electric market in China. In the U.S. and Europe, the ratio approached a 50:50 split. All-electric cars oversold plug-in hybrids in 2016, with pure electrics representing about 61% of the global stock, up from 58.9% at the end of 2015. The global ratio between battery BEVs and PHEVs was 66:34 in 2017. Cumulative global sales of highway-capable light-duty pure electric vehicles since 2010 achieved the one million unit milestone in September 2016.

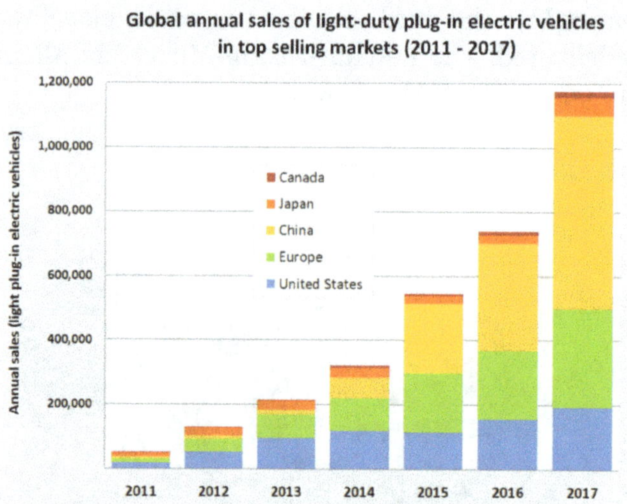

Annual sales of light-duty plug-in electric vehicles in the world's top markets between 2011 and 2017.

In five years, global sales of highway legal light-duty plug-in electric vehicles have increased more than ten-fold, totaling more than 565,000 units in 2015. Plug-in sales in 2015 increased about 80% from 2014, driven mainly by China and Europe. Both markets passed in 2015 the U.S. as the largest plug-in electric car markets in terms of total annual sales, with China ranking as the world's best-selling plug-in electric passenger car country market in 2015. About 775,000 plug-in cars and vans were sold in 2016, and cumulative global sales totaled an estimated 2.032 million plug-in cars and utility vans by the end of 2016. The global market share of the light-duty plug-in vehicle segment achieved a record 0.86% of total new car sales in 2016, up from 0.62% in 2015 and 0.38% in 2014. Despite the rapid growth experienced, the plug-in electric car segment represented just 0.15% of the 1.4 billion motor vehicles on the world's roads by the end of 2016, up from 0.1% in 2015. Global light-duty plug-in vehicle sales passed the 3 million milestone in November 2017. About 1.2 million plug-ins cars and vans were sold worldwide in 2017, with China accounting for about half of global sales in 2017.

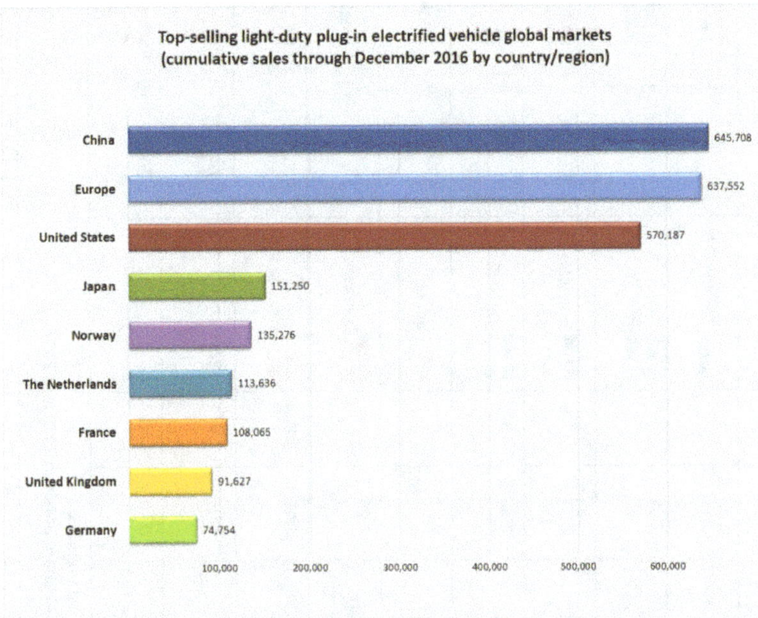

Top-selling light-duty plug-in electric vehicle global markets by country or region as of December 2016.

As of December 2016, with cumulative sales of more than 645,000 plug-in electric passenger cars, China has the world's largest fleet of light-duty plug-in electric vehicles, after having overtook during 2016 both the U.S. and Europe in terms of cumulative sales. This figure accounts for both, domestically produced new energy passenger cars and imports. The fleet of Chinese plug-in cars represents 43.0% of the global stock of light-duty plug-in vehicles at the end of 2016. Among country markets, the U.S. ranks second, with over 570,000 plug-in passenger cars sold through December 2016, representing 28.1% of the global stock of plug-ins. Japan has the world's third largest plug-in stock, with about 147,500 highway legal plug-in electric vehicles sold in the country between July 2009 and December 2016. As of December 2016, about 637,500 plug-in electric passenger cars and vans have been registered in Europe, representing 31.4% of the global stock, the second largest after China. The region was the world's largest light-duty plug-in market until October 2016.

As of December 2016, more than 951,000 new energy vehicles have been sold in China since 2011, making the country the world's leader when all plug-in automotive segments are considered, including passenger cars, electric buses, and commercial plug-in heavy-duty trucks. The stock of new energy vehicles in China achieved the 500,000 unit milestone in March 2016. As of December 2016, China had the world's largest electric bus stock with about 343,500 vehicles, up from 173,000 plug-in electric buses in 2015, both representing almost the entire global stock of plug-in buses. The European market account for 1,273 plug-in electric buses and the U.S. for only 200.

Top 10 countries or autonomous territories by plug-in electric passenger car market share of total new car sales between 2013 and 2017

Rank	Country or territory	Market share (%) 2017	Rank	Country or territory	Market share (%) 2016	Rank	Country or territory	Market share (%) 2015	Rank	Country or territory	Market share (%) 2014	Rank	Country or territory	Market share (%) 2013
1	Norway	39.2%	1	Norway	29.1 %	1	Norway	22.39 %	1	Norway	13.84 %	1	Norway	6.10 %
2	Iceland	14%	2	Netherlands	6.4 %	2	Netherlands	9.74 %	2	Netherlands	3.87 %	2	Netherlands	5.55 %
3	Sweden	5.2%	3	Hong Kong	~5 %	3	Hong Kong	4.84 %	3	Iceland	2.71 %	3	Iceland	0.94 %
4	Netherlands	2.6%	4	Iceland	4.6 %	4	Iceland	2.93 %	4	Estonia	1.57 %	4	Japan	0.91 %
5	Switzerland	2.6%	5	Sweden	3.5 %	5	Sweden	2.62 %	5	Sweden	1.53 %	5	France[2]	0.83 %
6	Belgium	2.6 %	6	Switzerland	1.8 %	6	Denmark	2.29 %	6	Japan	1.06 %	6	Estonia	0.73 %
7	Finland	2.6 %	7	Belgium	1.8 %	7	Switzerland	1.98 %	7	Denmark	0.88 %	7	Sweden	0.71 %
8	China	2.1%	8	Austria	1.6 %	8	France	1.19 %	8	Switzerland	0.75 %	8	US	0.60 %
9	Austria	2.1%	9	France	1.4 %	9	UK	1.07 %	9	US	0.72 %	9	Switzerland	0.44 %
10	France	1.98%	10	UK	1.37 %	10	Austria	0.90 %	10	France[2]	0.70 %	10	Denmark	0.29 %
Selected regional markets Plug-in electric passenger car market share between 2013 and 2017														
	California	4.8% %		California	3.6 %		California	3.1 %		California	3.2 %		California	2.5 %
	Europe	2%		Europe[1]	1.3 %		Europe[1]	1.41 %		Europe[1]	0.66 %		Europe[1]	0.49 %

(1) European figures correspond to European Union member countries plus EFTA countries (Norway and Switzerland)

(2) The French market share corresponds to combined sales all-electric passenger cars and utility vans only (plug-in hybrids not included).

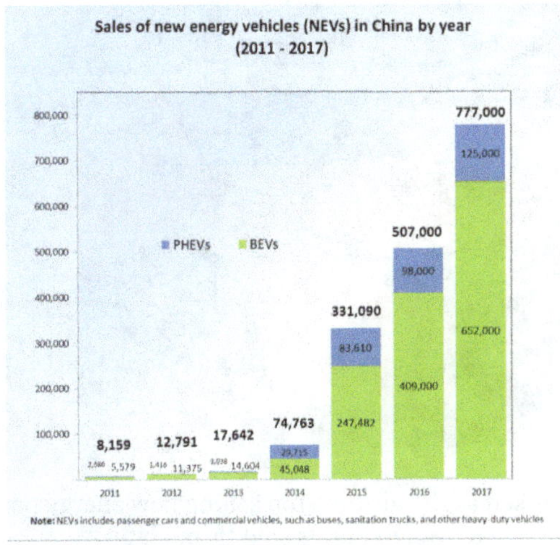

Annual sales of domestically built new energy vehicles in China by year (2011-2017).

Domestically built new energy vehicle sales in China totaled 1,728,447 units between 2011 and 2017. These figures include heavy-duty commercial vehicles such buses and sanitation trucks, and only include vehicles manufactured in the country as imports are not subject to government subsidies. China is the world's largest electric bus market, and its electric bus stock grew nearly sixfold between 2014 and 2015, reaching 343,500 units in 2016, doubling the 2015 stock.

As of December 2016, China ranked as the world's largest plug-in electric bus market with a stock of 343,500 vehicles. Shown a BYD K9 bus in Shenzhen.

As of December 2017, cumulative sales of domestically produced highway legal plug-in electric passenger cars totaled over 1,2 million units since 2005, of which, a total of 579,000 were sold in 2017, representing about half of global plug-in car sales in 2017. Domestically produced cars account for about 96% of new energy car sales in China. A particular feature of the Chinese passenger plug-in market is the dominance of small entry level vehicles.

In September 2016, the Chinese stock of plug-in passenger cars reached the same level of the American stock, and by November 2016, China's cumulative total plug-in passenger vehicles sales had surpassed those of Europe, allowing China to become the market with the world's largest stock of light-duty plug-in electric vehicles, with almost 600,000 plug-in passenger cars. China also surpassed the U.S. and Europe in terms of annual sales of light-duty plug-in electric vehicles, both in calendar years 2015 and current-year-to-date through November.

Since 2014 the BYD Qin ranked as the all-time top selling new energy passenger vehicle in China, with 89,393 units sold through 2017.

The stock of new energy vehicles sold in China since 2011 achieved the 500,000 unit milestone in March 2016, making the country the world's leader in the plug-in heavy-duty segment, including electric buses and plug-in trucks. Cumulative sales of new energy passenger cars achieved the 500,000 unit milestone in September 2016, excluding imports. As of December 2017, the BYD Qin plug-in hybrid, with 89,393 units sold since its inception, continued to rank as the all-time top selling plug-in electric car in the country. The BAIC EC-Series all-electric city car was Chinese the top selling plug-in car in 2017 with 78,079 units sold in China, making also the city car the world's top selling plug-in car in 2017. BYD Auto was again the top selling Chinese car manufacturer in 2017.

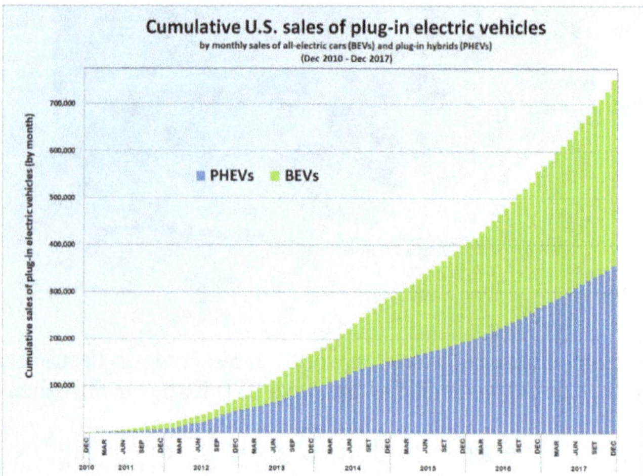

U.S. plug-in electric vehicle cumulative sales since 2008 passed the 500,000 unit milestone in August 2016. Graph shows cumulative sales by month by type of powertrain from December 2010 up to December 2017.

As of December 2017, cumulative sales totaled 764,666 highway legal plug-in electric cars in the U.S. since the market launch of the Tesla Roadster in 2008. As of December 2016, the American stock represented 28.1% of the global light-duty plug-in stock, down from about 40% in 2014. As of December 2016, the U.S. has the world's third largest stock of plug-in passenger cars, after having being overtook by Europe in 2015 and China during 2016. California is the largest U.S. car market, and accounts for approximately 48% percent of cumulative plug-in sales in the American market from 2011 to June 2016, and also accounts for about 50% of nationwide all-electric car sales and 47% of total plug-in hybrid sales. The other nine states that follow California's Zero Emission Vehicle

(ZEV) regulations have accounted for another 10% of cumulative plug-in car sales in the U.S. during the same period. California's plug-in stock totaled almost 270,000 units at the end of 2016.

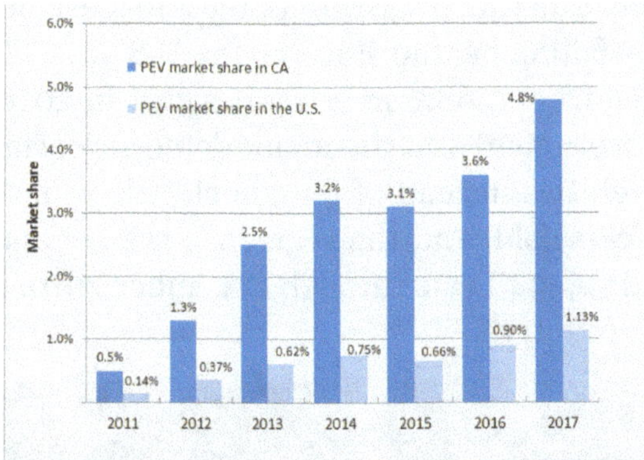

Comparison of plug-in electric car market shares California vs United States (2011-2017)

Nationwide sales climbed from 17,800 units delivered in 2011 to 53,200 during 2012, and reached 97,100 in 2013. In 2014 plug-in electric car sales totaled 123,248 units, and fell to 114,248 in 2015. A total of 157,181 plug-in cars were sold in 2016, up 37.6% from 2015, and rose to 194,479 in 2017, up 23.7% from 2016. The market share of plug-in electric passenger cars increased from 0.14% of new car sales in 2011 to 0.37% in 2012, 0.62% in 2013, and reached 0.75% of new car sales in 2014. As plug-in car sales slowed down during 2015, the segment's market share fell to 0.66% of new car sales, but increased to 0.90% in 2016. The plug-in segment passed the 1% market share for the first time in 2017, with 1.13% of the country's total annual new car sales. The highest-ever monthly market share for plug-in electric vehicles was achieved in December 2017 with 1.58% of new car sales. The previous record was achieved in September 2017. Monthly sales of plug-in cars in the American market passed the 1% mark for the first time in September 2016 (1.12%). December 2017 is also the best monthly plug-in sales volume on record ever, with 25,149 units delivered.

f December 2016, the Nissan Leaf is the all-time top selling plug-in electric car in Japan, with almost 72,500 units sold since inception.

apan had a stock of light-duty plug-in vehicles of about 207,200. As of onsisted of 86,390 all-electric cars (57.1%) and 64,860 plug-in hybrids les climbed from 1,080 units in 2009 to 12,630 in 2011, and reached f pure electric cars in 2012 were led by Japan with a 28% market ranked second after the U.S. in terms of its share of plug-in hy- the country ranked third.

The plug-in segment sales remained flat in 2014 with 30,390 units sold, and a market share of 1.06% of total new car sales in the country (kei cars not included). Sales totaled 24,660 units in 2015 and 24,851 units in 2016. The rate of growth of the Japanese plug-in segment slowed down from 2013, with annual sales falling behind Europe, the U.S. and China during 2014 and 2015. The segment market share fell from 0.68% in 2014 to 0.59% in 2016. The decline in plug-in car sales reflects the Japanese government and the major domestic carmakers decision to adopt and promote hydrogen fuel cell vehicles instead of plug-in electric vehicles. Sales recovered in 2017, with almost 56,000 plug-in cars sold, and the segment's market share reached 1.1%. Since December 2010, Nissan has sold 72,494 Leafs through December 2016, making the Leaf the all-time best-selling plug-in car in the country.

The Renault Zoe was the top selling plug-in electric car in Europe in 2016 and 2017.

As of December 2017, about 943,600 plug-in electric passenger cars and vans have been registered in Europe, making the region the world's second largest after China. As of December 2017, European sales of plug-in cars and vans are led by Norway with over 200,000 units registered, followed by France with almost 158,000, and the UK with over 137,000 units. Norway was the top selling plug-in country market in Europe in 2016 and 2017. The other top selling European markets in terms of cumulative registrations are Germany, the Netherlands and Sweden.

Cumulative sales of light-duty plug-in electric vehicles in Europe passed the 500,000 unit milestone in May 2016. Norway passed the 100,000th registered plug-in unit milestone in April 2016, France passed the same milestone in September 2016, and the Netherlands in November 2016. The UK achieved the 100,000 unit mark in March 2017.

As of December 2015, France ranked as the largest European market for light-duty electric commercial vehicles or utility vans, accounting for nearly half of all vans sold in the European Union. The French market share of all-electric utility vans reached a market share of 1.22% of new vans registered in 2014, and 1.30% in 2015. Denmark is the second largest European market, with over 2,600 plug-in electric vans sold in 2015, with an 8.5% market share of all vans sold in the country. Most of the van sold in the Danish market are plug-in hybrids, accounting for almost all of th[illegible] plug-in hybrid van sales across the EU. European sales of plug-in electric cars passed 2[illegible] units for the first time in 2016, and the plug-in segment achieved a market share of [illegible] 2017 exceeded 300,000 units with a market share of 2%.

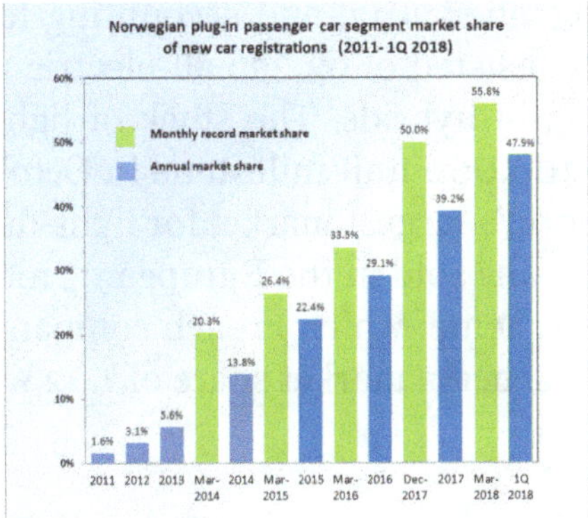

Historical evolution of the Norwegian plug-in electric car segment market share of new car sales and monthly records through March 2018.

As of December 2017, the stock of light-duty plug-in electric vehicles registered in Norway totaled more than 200,000 units, making Norway the European country with the largest stock of plug-in cars and vans, and the fourth largest in the world. As of 31 December 2017, and accounting for both new and used imports registrations, the Norwegian light-duty plug-in electric fleet consisted of 141,951 all-electric passenger cars and vans, and 67,171 plug-in hybrids. The registered plug-in stock includes almost 2,700 all-electric vans and about 24,500 used imported electric cars from neighboring countries as of December 2017. The government's target of 50,000 all-electric cars on Norwegian roads was reached in April 2015, more than two years earlier than expected, thanks to the successful policies implemented to promote electric vehicle adoption that include fiscal and non-monetary incentives. The milestone of 100,000 light-duty plug-in electric vehicles registered was achieved in April 2016, and 100,000 all-electric vehicles in December 2016, representing about 10% of all pure electric cars that have been sold worldwide.

The Norwegian fleet of plug-in electric cars is one of the cleanest in the world because 98% of the electricity generated in the country comes from hydropower. Norway, with about 5.2 million people, is the country with the largest EV ownership per capita in the world. In March 2014, Norway became the first country where over one in every 100 registered passenger cars is plug-in electric. The plug-in car market penetration reached 2% in March 2015, passed 3% in December 2015, and achieved 5% at the end of 2016.

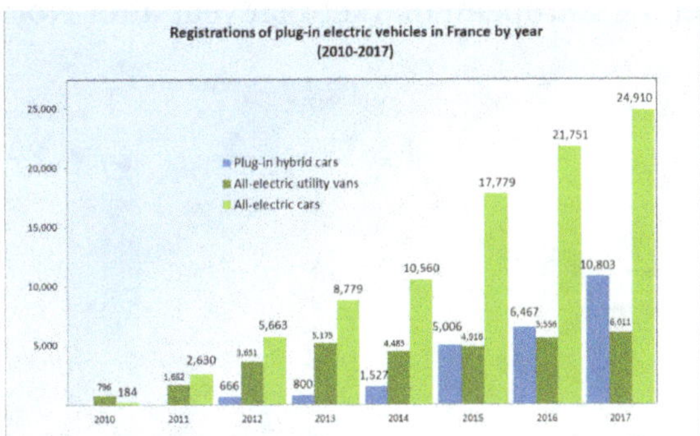

Registration of highway capable plug-in electric vehicles in France by type of vehicle between 2010 and 2017.

As of December 2017, the stock of light-duty plug-in electric vehicles registered in France totaled 149,797 plug-in cars and electric utility vans delivered since 2010, As of December 2016, the country ranked as the third largest plug-in market in Europe after Norway and the Netherlands, and the world's sixth. As of December 2017, and accounting for registrations since 2010, the plug-in electric stock consisted consisted of 92,256 all-electric passenger cars, 25,269 all-electric utility vans, and 32,272 plug-in hybrids. The stock of light-duty plug-in electric vehicles registered in France passed the 100,000 unit milestone in October 2016.As of December 2015, France is the country with the world's largest market for light-duty electric commercial vehicles or utility vans. Nearly half of the vans sold in the European Union are sold in France as a result of a national purchase incentive scheme, which French companies have embraced. The market share of all-electric utility vans reached a market share of 1.22% of new vans registered in 2014, and 1.30% in 2015.

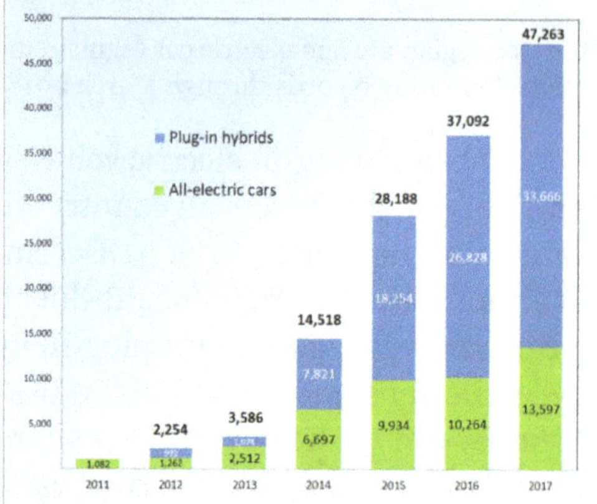

Registration of plug-in electric vehicles in the UK between 2011 and December 2017.

More than 137,000 light-duty plug-in electric vehicles have been registered in the UK up until December 2017, including about 5,100 plug-in commercial vans. Since the launch of the Plug-In Car Grant in January 2011, a total of 127,509 eligible cars have been registered through December 2017, and, as of December 2016, the number of claims made through the Plug-in Van Grant scheme totaled 2,938 units since the launch of the scheme in 2012. Before the introduction of series production plug-in vehicles, a total of 1,096 all-electric vehicles were registered in the UK between 2006 and December 2010. Before 2011, the G-Wiz, a heavy quadricycle, listed as the top-selling EV for several years. As of December 2016, the UK had the fourth largest European stock of light-duty plug-in vehicles, and with 36,907 plug-in passenger cars registered in 2016, ranked as the second best selling European market that year after Norway.

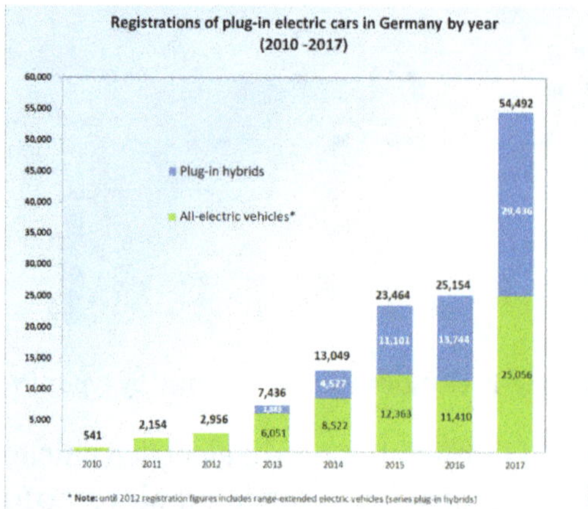

Annual registration of plug-in electric cars in Germany by type of vehicle between 2010 and 2017.

As of December 2017, a total of 129,246 plug-in electric cars have been registered in Germany since 2010. The country is the largest passenger car market in Europe, however, as of December 2016, ranked as the fifth largest plug-in market in Europe. About 80% of the plug-in cars registered in the country through September 2016 were registered since January 2014. The official German definition of electric vehicles changed at the beginning of 2013, before that, official statistics only registered all-electric vehicles because plug-in hybrids were accounted together with conventional hybrids. As a result, the registrations figures for 2012 and older do not account for total new plug-in electric car registrations.

A record of 54,492 plug-in cars were registered in 2017, up 217% the previous year, and consisting of 29,436 plug-in hybrids and 25,056 all-electric cars. Registrations in 2017 achieved a market share of 1.58%. The top selling models in 2017 were the Audi A3 e-tron (4,454), Renault Zoe (4,322), and BMW i3 (4,319).

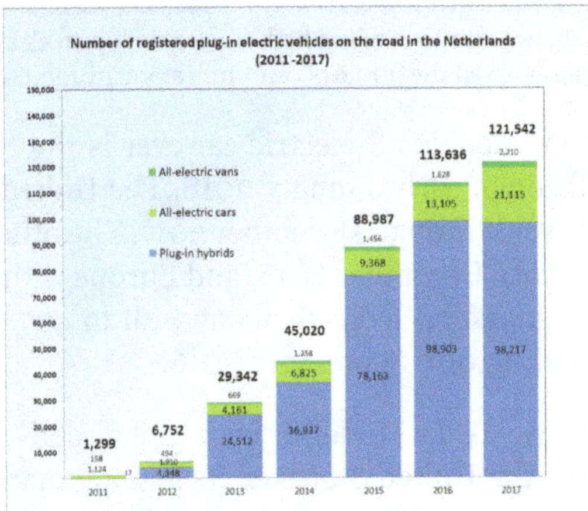

Number of registered plug-in electric vehicles on the road in the Netherlands from 2011 to 2017.

As of 31 December 2017, there were 121,542 highway legal light-duty plug-in electric vehicles registered in the Netherlands, consisting of 98,217 range-extended and plug-in hybrids, 21,115 pure electric cars, and 2,210 all-electric light utility vans. When buses, trucks, motorcycles, quadricycles and tricycles are accounted for, the Dutch plug-in electric-drive fleet climbs to 123,499 units. The country's electric vehicle stock reaches 165,886 units when fuel cell electric vehicles (43), mopeds (4,376), electric bicycles (37,652), and microcars (316) are accounted for. A distinct feature of the Dutch plug-in market is dominance of plug-in hybrids, which represented 81% of the country's stock of passenger plug-in electric cars and vans registered at the end of December 2017.

The Netherlands listed as the world's third best-selling country market for light-duty plug-in vehicles in 2015, with 43,971 units registered that year. Until December 2015, the Netherlands had Europe's largest fleet light-duty plug-in vehicles. Plug-in sales fell sharply during 2016, and as a result, by early October 2016, the Netherlands listed as the third largest European plug-in market, after being surpassed during the year by both Norway and France. As of July 2016, the Netherlands had the second largest plug-in market concentration per capita in the world after Norway. The stock of light-duty plug-in electric vehicles registered in the Netherlands achieved the 100,000 unit milestone in November 2016. However, due to changes in tax rules, the plug-in market share declined from 9.9% in 2015, to 6.7% in 2016, and fell to 2.6% in 2017.

Top Selling PEV Models

All-electric Cars and Vans

The Nissan Leaf is the world's top selling highway legal plug-in electric car in history. Global sales passed the 300,000 unit milestone in January 2018.

The world's top selling highway-capable all-electric car ever is the Nissan Leaf with global sales of more than 300,000 units through early January 2018. The United States is the world's largest Leaf market with 114,827 units sold through December 2017. The other two top markets are Japan with 74,494 units delivered up until December 2016, and Europe with about 66,000 units through November 2016. The European market is led by Norway with 19,407 new units registered up until December 2016, and the UK with 15,000 September 2016.

Ranking second is the all-electric Tesla Model S, with global deliveries of more than 200,000 cars as of November 2017, with the United States as its leading market with about 118,817 units delivered through December 2017. As of September 2016, Norway ranked as the Model S largest overseas market, with 11,763 new units registered, followed by China with 5,524 units registered through September 2015. The Tesla Model S was the world's top selling plug-in car for two years in a row, 2015 and 2016. The BAIC EC-Series all-electric city car ranked as the world's top selling plug-in car in 2017 with 78,079 units sold in China. The world's all-time top selling all-electric light utility vehicle is the Renault Kangoo Z.E., with global sales of 29,523 electric vans delivered since inception through December 2017.

The following table presents global sales of the top selling highway-capable electric cars and light utility vehicles produced between 2008 and December 2016. The table includes all-electric passenger cars and utility vans with cumulative sales of about or over 20,000 units since the introduction of the first modern production all-electric car in 2008, the Tesla Roadster.

Top selling highway legal electric cars and light utility vehicles produced between 2008 and December 2016[(1)]			
Model	Market launch	Global sales	Sales through
Nissan Leaf	Dec 2010	+ 250,000	Dec 2016
Tesla Model S	Jun 2012	158,159	Dec 2016
BMW i3	Nov 2013	~ 65,500	Dec 2016
Renault Zoe	Dec 2012	61,205	Dec 2016

BAIC EV series	2012	42,626[(3)]	Dec 2016
Mitsubishi i-MiEV family	Jul 2009	~ 37,600	Jun 2016
BYD e6	Oct 2011	34,862[(3)]	Dec 2016
Tesla Model X	Sep 2015	25,524	Dec 2016
Renault Kangoo Z.E.	Oct 2011	25,205	Dec 2016.
Volkswagen e-Golf	May 2014	24,498[(4)]	Jun 2016
JAC J3/iEV family	2010	23,241[(3)]	Jun 2016
Cumulative global sales 2008 – Dec 2016		over 1,230,000	Dec 2016
(1) Vehicles are considered highway-capable if able to achieve at least a top speed of 100 km/h (62 mph). Several models, such as the Chery QQ3 EV/eQ EV, Kandi EV and the Zotye Zhidou E20, are highway legal in China but do not meet this requirement. (2) BMW i3 sales includes the REx variant. (3) Sales in main China only. (4) Sales in Europe and the U.S. only.			

Plug-in Hybrids

The Volt/Ampera family is the world's best selling plug-in hybrid. Combined global sales of Volt and Ampera models passed the 100,000 unit milestone in October 2015.

As of December 2016, the Volt/Ampera family is the world's best selling plug-in hybrid and the third best selling plug-in electric car after the Nissan Leaf and the Model S. Chevrolet Volt and Opel/Vauxhaul Ampera combined sales totaled about 134,500 units worldwide through December 2016, including just over 10,000 Opel/Vauxhall Amperas sold in Europe through June 2016. Volt sales are led by the United States with 113,489 units delivered, followed by Canada with 8,884 units, both through December 2016. The Netherlands is the leading European market with about 6,000 Amperas and Volts registered as of December 2015.

Ranking next is the Mitsubishi Outlander P-HEV with around 119,500 units sold worldwide as of December 2016. As of March 2016, Europe ranked as the Outlander P-HEV leading market with 65,529 units sold, followed by Japan with 33,730 units. European sales are led by the UK with 26,600 units sold through December 2016, followed by the Netherlands with 25,984 units registered by the end of 2016.

Ranking third is the first generation Toyota Prius Plug-in Hybrid with 75,400 units sold worldwide through April 2016. The United States is the market leader with 44,767 units delivered through December 2016. Japan ranks next with about 22,100 units, followed by Europe with 10,500 units, both, through April 2016. The leading European market is the Netherlands with 4,134 units registered as of 30 November 2015. Production of the first generation Prius Plug-in ended in June

2015. The second generation Prius plug-in hybrid, the Toyota Prius Prime, was released in the United States in November 2016..The following table presents cumulative sales through December 2016 of those plug-in hybrid models that have sold about 10,000 units since the introduction of the first modern production plug-in hybrid vehicle in December 2008, the BYD F3DM.

Top selling highway legal plug-in hybrid electric cars between 2008 and December 2016			
Model	Market launch	Global sales	Sales through
Chevrolet Volt(1)	Dec 2010	~134,500	Dec 2016
Mitsubishi Outlander P-HEV	Jan 2013	~119,500	Dec 2016
Toyota Prius PHV	Jan 2012	~77,850	Dec 2016
BYD Qin(2)	Dec 2013	68,655	Dec 2016
BYD Tang(2)	Jun 2015	49,780	Dec 2016
Ford Fusion Energi(3)	Feb 2013	43,713	Dec 2016
Ford C-Max Energi(4)	Oct 2012	35,705	Dec 2016
SAIC Roewe 550 PHEV(2)	Nov 2013	26,856	Dec 2016
Volkswagen Golf GTE	Aug 2014	24,089	Jun 2016
Volvo V60 Plug-in Hybrid	Nov 2012	21,185	Jun 2016
Audi A3 Sportback e-tron(5)	Aug 2014	18,467	Jun 2016
BMW i8	Aug 2014	10,000	Nov 2016
Volvo XC90 T8(5)	Aug 2015	9,044	Jun 2016
Cumulative global sales Dec 2008 – Dec 2016		~800,000	Dec 2016

(1) Includes Vauxhall/Opel Ampera sales in Europe, and Holden Volt in Australia and New Zealand. This figure includes over 10,000 Opel/Vauxhall Amperas sold in Europe through December 2016.
(2) Sales in China only.
(3) Sales in the U.S. and Canada only.
(4) Only includes European sales for 2015.
(5) Only accounts for CYTD 2016 sales in Europe, U.S. and Canada.

Battery Electric Vehicle

Battery electric vehicles, or BEVs, use electricity stored in a battery pack to power an electric motor and turn the wheels. When depleted, the batteries are recharged using grid electricity, either from a wall socket or a dedicated charging unit. Since they don't run on gasoline or diesel and are powered entirely by electricity, battery electric cars and trucks are considered "all-electric" vehicles.

When driven, BEVs don't produce tailpipe pollution—they don't even have a tailpipe. However, the electricity they use may produce heat-trapping gases and other pollution at the source of its generation or in the extraction of fossil fuels. The amount of pollution produced depends on how the electricity is made. In the United States, battery electric cars charged off the dirtiest coal-dominated grid *still* produce less pollution than their gasoline-powered counterparts. BEVs powered by renewable energy sources like wind or solar are virtually emission-free.

Not using gasoline or diesel also means that battery electric cars are significantly cheaper to fuel than conventional vehicles. Exact comparisons depend on the vehicle model and fuel prices, but driving a BEV can save drivers over $1,000 annually in gasoline money.

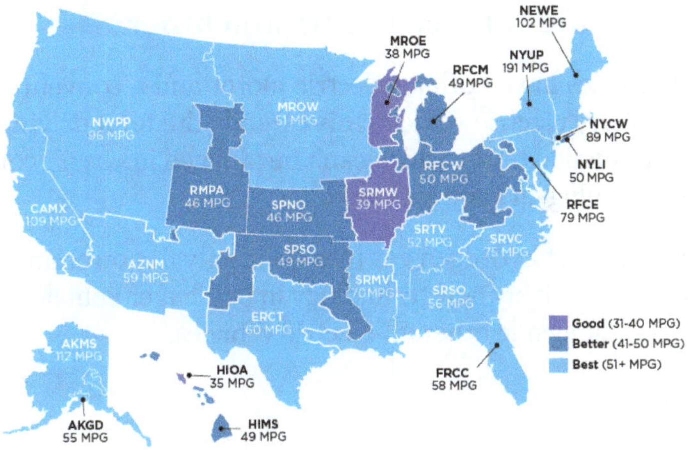

2009-2016

Battery Electric Vehicle Features

Like other electric and hybrid-electric vehicles, BEVs minimize wasted energy by turning the car off when stopped ("idle-off") and by charging the battery when braking ("regenerative braking"). Electric motors are also inherently more energy-efficient than gasoline or diesel engines.

Battery electric cars run entirely on electricity and can be significantly cleaner than gas-powered vehicles.

Battery electric cars have the added benefit of home recharging. A 240-volt outlet, similar to those used for clothes dryers, can charge a vehicle overnight. Fully-charged, most battery electric cars have a driving range of between 70 to 100 miles, well within the day-to-day range requirements of most Americans, though some BEVs can go up to 265 miles on a single charge. An increasing number of public and workplace charging stations provide added charging capacity.

More subjectively, many drivers appreciate the driving experience provided by all-electric cars. Electric motors generate near-instant torque, or turning-force, while the torque of internal combustion engines increases in tandem with the engine's revolutions (RPM). This means that BEVs have extremely fast acceleration and a "light" or "zippy" feel compared to conventional cars and trucks.

Differences between Battery Electric Cars and other EVs

Plug-in hybrid electric vehicles have both an electric motor and a conventional gasoline or diesel engine. Compared to a battery electric vehicle, this extends the total driving range but lowers the all-electric range. Conventional hybrids, which can't be plugged in, aren't considered electric vehicles. Learn more about how plug-in vehicles work.

Fuel cell electric vehicles convert hydrogen gas into electricity to power an electric motor and battery. Fuel cell vehicles are a relatively new technology in passenger vehicles, but have a substantial carbon-cutting role to play alongside other all-electric vehicles.

Hybrid Electric Vehicles

A hybrid vehicle combines any two power (energy) sources. Possible combinations include diesel/electric, gasoline/fly wheel, and fuel cell (FC)/battery. Typically, one energy source is storage, and the other is conversion of a fuel to energy. The combination of two power sources may support two separate propulsion systems. Thus to be a True hybrid, the vehicle must have at least two modes of propulsion.

For example, a truck that uses a diesel to drive a generator, which in turn drives several electrical motors for all-wheel drive, is *not a hybrid.* But if the truck has electrical energy storage to provide a second mode, which is electrical assists, then it is a hybrid Vehicle.

These two power sources may be paired in series, meaning that the gas engine charges the batteries of an electric motor that powers the car, or in parallel, with both mechanisms driving the car directly.

Hybrid Electric Vehicle (HEV)

Consistent with the definition of hybrid above, the hybrid electric vehicle combines a gasoline engine with an electric motor. An alternate arrangement is a diesel engine and an electric motor.

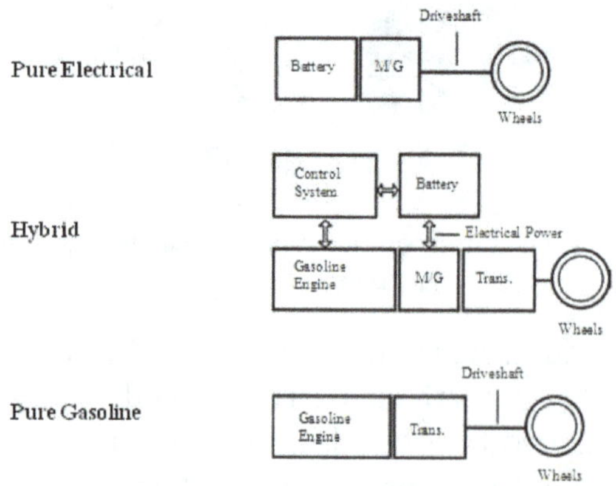

Figure: Components of a hybrid Vehicle that combines a pure gasoline with a pure EV.

As shown in Figure, a HEV is formed by merging components from a pure electrical vehicle and a pure gasoline vehicle. The Electric Vehicle (EV) has an M/G which allows regenerative braking for an EV; the M/G installed in the HEV enables regenerative braking. For the HEV, the M/G is tucked directly behind the engine. In Honda hybrids, the M/G is connected directly to the engine.

The transmission appears next in line. This arrangement has two torque producers; the M/G in motor mode, M-mode, and the gasoline engine. The battery and M/G are connected electrically.

HEVs are a combination of electrical and mechanical components. Three main sources of electricity for hybrids are batteries, FCs, and capacitors. Each device has a low cell voltage, and, hence, requires many cells in series to obtain the voltage demanded by an HEV. Difference in the source of Energy can be explained as:

- The FC provides high energy but low power.
- The battery supplies both modest power and energy.
- The capacitor supplies very large power but low energy.

The components of an electrochemical cell include anode, cathode, and electrolyte. The current flow both internal and external to the cell is used to describe the current loop.

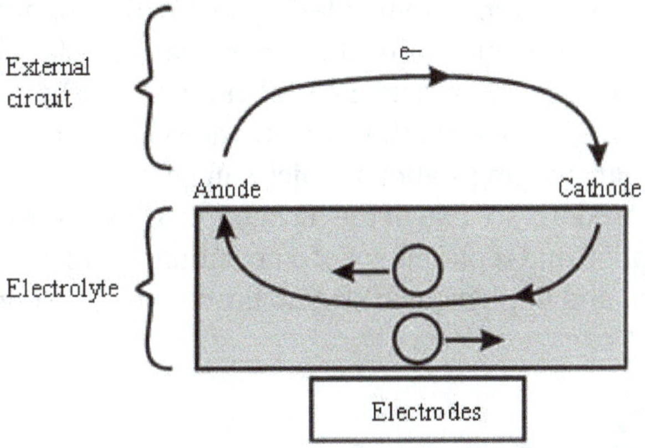

An electrode, a circuit for a cell which is converting chemical energy to electrical energy. The motion of negative charges is clockwise and forms a closed loop through external wires and load and the electrolyte in the cell.

A critical issue for both battery life and safety is the precision control of the Charge/Discharge cycle. Overcharging can be traced as a cause of fire and failure.

Applications impose two boundaries or limitations on batteries. The first limit, which is dictated by battery life, is the minimum allowed State of Charge. As a result, not all the installed battery energy can be used. The battery feeds energy to other electrical equipment, which is usually the inverter. This equipment can use a broad range of input voltage, but cannot accept a low voltage. The second limit is the minimum voltage allowed from the battery.

Present of Hybrid Electric Vehicle

Toyota is the most prominent of all manufacturers when it comes to hybrid cars. As well as the specialist hybrid range they have produced hybrid versions of many of their existing model lines, including several Lexus (now owned and manufactured by Toyota) vehicles. They have also stated that it is their intention to release a hybrid version of every single model they release in the coming decade. As well as cars and SUVs, there are a select number of hybrid motorcycles, pickups, vans, and other road going vehicles available to the consumer and the list is continually increasing.

Future of Hybrid Electrical Vehicle

Since petroleum is limited and will someday run out of supply. In the arbitrary year 2037, an estimated one billion petroleum-fueled vehicles will be on the world's roads. Gasoline will become prohibitively expensive. The world need to have solutions for the *"400 million otherwise useless cars"*. So year 2037 "gasoline runs out year" means, petroleum will no longer be used for personal mobility. A market may develop for solar-powered EVs of the size of a scooter or golf cart. Since hybrid technology applies to heavy vehicles, hybrid buses and hybrid trains will be more significant.

Electric Car

Concerns over rising fuel prices, depleting resources and the desire to lower greenhouse gas emissions, have necessitated innovation in automobiles or reinvention of automobile technology. One such innovation has been the development of the electric car. An electric car is a PEV or plug-in electric vehicle, which is propelled by electric motors and powered by rechargeable batteries. It is much quieter compared to the internal combustion engine vehicle and has lower or no emissions. Electric cars can be charged at charging stations, which can be installed in private and public spaces. The Nissan Leaf and Tesla Model S are two of the most popular hybrid electric vehicles in the world today. To promote the adoption and replacement of conventional vehicles with modern automobiles, governments have proposed and implemented several tax credits, subsidies and incentives.

Electric Bikes

Electric Bikes consist of a modified or custom bicycle frame with pedals but include an electric motor, usually in the form of a hub motor, mid-drive motor or belt drive connected to the rear wheel. They allow a rider to either pedal the bike or leverage the power of a battery and motor drive system.

To use an electric bike you have to charge the battery, turn the controller system on and then twist a throttle, pull a trigger or pedal the bike forward to initiate the motor drive system. Throttle only drive behaves much like a traditional gas powered scooter because the rider doesn't have to pedal, the bike just goes! By contrast, pedal-assist mode uses the rider's input as the throttle signal and is more popular in Europe where bicycle laws are a bit more restrictive regarding motor and battery power. European bikes are usually limited to 250 Watt motors whereas the US has many bikes with motors as strong as 500 or even 600w. These bikes go faster and can power up steeper hills more easily.

Advantages of Owning an E-bike

- Improved Fitness

  People that own E-bikes have improved fitness because they use them more frequently and ride for greater distances than on a conventional bike. Many people say e-bikes are cheating, but if it gets you out riding, there is nothing wrong with them. Any activity added to your day is always beneficial even moderate amounts. Plus depending on your energy level you can rely less on power and do more peddling.

- No Sweat

  A big issue with commuting to work in nice clothing is that you arrive at work all sweaty. An E-bike can solve this problem by allowing you to peddle less and using more assistance power when going to work. You can have a change of clothing, peddle more and use less battery power for more of a workout on the way home.

- Helps in Challenging Conditions

  Makes biking much less challenging when cycling into headwinds and going up hills.

- More Acceleration then Convention Bikes

  An E-Bike gives you more acceleration then convention bikes and can be safer for driving in busy traffic. This extra acceleration can really help when maneuvering around cars and get you where you are going despite the traffic.

- Helps Seniors and People with Physical Limitations

  Great for a broad range of people that could use little assistants and can benefit from cycling. People that otherwise would have to give up cycling due to physical limitations can benefit from an E-bike. People with COPD or asthma, bad knees or have heart conditions all can benefit by using an E-bike.

  For a senior, an E-bike can be just the thing you need to maintain your independence. An E-bike can keep you riding longer and give you a little boost when you need it.

- Easier to Carry Heavier Loads

  An E-Bike can make it easier to carry heavier loads when shopping and make it easier to have a passenger on board or tow a trailer for children.

- People that Own E-bikes will Commute Further

  People that own E-bikes will commute further to work than on a conventional bike. With improved bike paths and more consideration being given to bikes implemented in many cities make travel by bike even more enjoyable.

  Some E-bikes are small enough or can be folded to take on a train or bus. As with many new bike paths, many cities have added bike racks on city buses. This can make transporting your bike to another area of town easy One caveat is an E-bike can be very heavy so loading your bike on a bus rack could be a little difficult for some people.

- Easy to Park and no Parking Fees

  No more circling to find a parking spot. You can easily take it on an elevator and park it inside for maximum theft safety and storage. Bike racks are becoming more common and are often located even closer to the door than handicap parking.

- No Registration, Insurance or License is Required

  No registration, insurance or license is required to ride an E-bike, provided you are not exceeding the power and speed rating classification in your area.

- Canada maximum 500-watt power and go faster than 20 mph / 32 km/h
- US maximum 750-watt power and go faster than 20 mph / 32 km/h
- E-bike are very Inexpensive to Run and can be Charge Anywhere

  An E-bike gives you a quite enjoyable ride that gives a little extra boost when you need it. It is environmental friendly and can be charged from any standard outlet. An E-bike can be very inexpensive to run. Example to fully charge a battery that will go a range of 30 to 50 miles cost around 10 cents. Much cheaper than gas.

- E-bikes have Low Maintenance Costs

  Compared to a car you will have much lower maintenance costs on an E-bike. As you know any repairs on an automobile can easily jump up to $500 or more and that may not be all the problems. With a $500 budget you can get a major overhaul done on a bike.

There are many advantages to riding an E-bike but what seems to be the biggest advantage is that it is simply fun to ride. You can ride to work with minimal stress and an invigorating ride can really improve your daily mood. It can be a really enjoyable way to commute. Traffic is almost irrelevant when you're riding a bike you will never be stuck in traffic and always be moving along at a moderate speed. Unlike sitting frustrated stuck in a car not moving.

Disadvantages of Owning an E-bike

As with everything there are always a few disadvantages to owning an E-bike that you may want to consider before going out and buying one.

- High Cost of E-bikes

  This is the main beef with many people. That E-bikes are too expensive. I would compare the new E-bike market to the plasma TV market twenty years ago. When flat screen TVs first came out they sold for four or five thousand dollars and a few years later there were many being offered for under $2000. Today you can get a big flat screen for under $1000.

I think the same is happening with E-bikes. Prices are coming down and more E-bikes are being offered for under $2000 and I believe more really good E-bikes are coming in this price range as demand for E-bikes grows.

If you put it into perspective and add up the saving over a few years, E-bikes are not so bad a deal. Consider the cost saving over a few years without having to pay for registration, licensing, insurance, and parking and let's not forget gas. Add this up and the monthly saving pays for the bike in a few years. Plus you might loss a few pounds.

- More Expensive to Run than Conventional Bike

  Not only is the initial cost of an e-bike more than a conventional bike the ongoing maintenance and repair costs can cost much more. E-bikes have three expensive systems a normal bike does not have. The power hub, the battery and the controller all can be quite costly to replace, especially the battery.

  That is why you want to practice good battery charging habits to ensure you get the most out of your battery. E-bikes normally go a little faster and use power so there is a tendency to use your brakes more and you will have to be replaced them more often. E-bikes vibrate more and most people have a tendency to ride more on an E-bike. Because of this servicing is required more often on an E-bike.

- E-bikes are Heavier than Conventional Bike

  Many bikes over the years have been getting lighter and stronger made from aluminum or carbon fiber to produce a very light bike. Unfortunately, this is not the case with an E-bike. Batteries have been getting lighter by they still add a significant amount of weight to the overall weight of the bike. The hub motor is also no lightweight and further add to the bikes weight.

  Due to this extra weight, an E-bike has to be extra strong to carry you and the weight of the bike and some of the lighter materials being used on conventional bikes will not work well with an E-bike.

  This can be a problem for some people if they have to carry a bike up stairs or lift it onto a bike rack on a bus. Taking the battery off when lifting can make it easier but sometimes this option is not practical.

- Recharging Battery takes Some Time

  Unlike a gas-powered engine were you just fill up and go a battery requires up to 3 to 4 hours to fully charge. This can be a problem with bad planning and you find yourself with a dead battery. But with proper planning and charging techniques, this problem can be greatly reduced. The good thing is you can still peddle home if you do run out of power.

- Theft is a Concern

  It would be a nice world if people would not steal your stuff. But unfortunately this is not the case. E-bikes cost more and have more expensive parts make E-bikes an attractive target for a thief. Many convenient features designed for easy maintenance are also ideal for thieves to quickly remove parts. The best thing to do is bring your bike inside if you can and if not buy good quality locks and lock up your bike properly.

- Limited Range

  Many E-bikes have a range of between 30 and 50 miles on a single charge. This limitation makes it important to plan your trip and figure out how far you have to go before running out of power and have to rely on peddle power alone.

  There are a number of companies like Tesla that are working on improving rechargeable batteries making them more powerful and lighter. I think we will see more improvement in battery life in the next few years.

- Exposed to the Elements

  Unless you live in southern California you will probably encounter rain and nasty weather on a number of occasions when you are commuting to work. Biking in the rain can be no fun and sometimes downright dangerous. Visibility is more limited to you and the other drivers. Operating a bike in the rain can be more difficult. Wet brakes don't work very well and traction can get very slippery especially at the start of a rain.

  If it is a washout of a day take a bus or get a ride. There are also a number of ponchos available that will cover you and your bike when riding in the rain.

  You don't have to worry about riding your E-bike in the rain. Most E-bikes are designed to take a little splashing without any problems. Just don't submerge your E-bike in water and don't wash your E-bike with a pressure washer.

- Not as much Exercise as Conventional Bike

  A number of people say that using an E-bike is cheating and you are not getting as much exercise. In some ways, they may be right. But if an E-bike actually gets you out riding whereas with a regular bike you would leave it at home than the E-bike wins for fitness. A little is better than nothing and you have the option to do more if you peddle harder and rely less on power.

- Battery Issues

  E-bike batteries cannot be taken on airlines because some lithium batteries have exploded in devices in the past. Many batteries are also made from toxic chemicals and cannot be recycled. Replacing a bad battery is also quite expensive.

- Lower Speed and more Vulnerable in Traffic

  Many drivers are not very considerate to bikes on the road especially in North America were bike transportation has always taken a back seat to the automobile. E-bikes don't go that fast so getting around in traffic can be more dangerous. Bad weather also makes it more difficult for drivers to see you. That is why being seen is very important for proper road safety.

- No Creature Comforts and Extra Conveniences

  Riding a bike is less comfortable than driving in a car. On hot days there is no AC and on cold days there is no heat. Answering a phone call or listening to music is not as convenient and can be much more dangerous on a bike.

Stopping off at the grocery store or anywhere for a number of items is not as easy. An E-bike has a limited load carrying capabilities. This can pose a problem depending what you need to bring back and forth to work.

Some cycle seats can be quite uncomfortable and can be a major factor in why someone does not like riding their bike.

Electric Bus

The electric bus is powered by energy stored in rechargeable batteries rather than fossil fuels. Instead of an internal combustion engine, it's propelled by an electric motor. The principle is similar to a series hybrid transit bus but rather than an engine powering an AC generator that supplies power to an AC electric drive motor, the batteries supply power to the motor. The batteries supply high voltage DC power that is inverted to 3 phase AC power with variable frequency control. The AC motor speed and therefore the speed of the bus is related to the frequency of the power supplied by the control system. The frequency is determined by the bus operator who uses the accelerator pedal to select the speed of the bus based on the required driving conditions.

Much of the braking force of the bus is also provided by the electric drive system. When the operator applies the brakes, the control system converts the electric drive motor into a generator that absorbs the kinetic energy of the moving bus to slow down the bus and transform that energy into power to recharge the batteries.

The lithium ion batteries store up to 200 kWh of electricity and are monitored by a sophisticated battery management system that ensures the batteries operate efficiently and that their lifespans are maximized.

Charging of the Bus

Buses will pull up to a specially constructed terminal at the airport that houses a rapid charging station. A roof mounted pantograph (similar to the technology used in city trams around the world) connects to the overhead charger interface. The pantograph is deployed automatically without the operator leaving the driver seat. Before charging begins, the system goes through a verification routine that ensures the bus is properly connected to the station and the charging process is fully automatic.

The charging station supplies up to 300 kW of power to the bus and has the capacity to recharge the bus in about 10 minutes after each round trip on the route. Charging will occur during the scheduled layover time at the airport terminal while the bus is waiting to start its next trip.

The bus will leave the airport for the outer terminal in East Kildonan through the downtown and will return to the airport approximately two hours later. The bus, which has the battery capacity to allow for operation of up to five hours between charges, is capable of maintaining service for more than 20 hours a day, seven days a week, when kept at full charge.

Electric buses need infrastructure, regulations and an entire ecosystem. There is substantial effort already ongoing in each of these areas. Key points being addressed for the charging infrastructure are standardising charging station interfaces, regulatory changes to decide on which entities can sell power, defining safety standards, and an entity that can enable the necessary space for installation of charging stations.

There are different approaches to charging infrastructure. I am an advocate of 'Top-up Charging' —one that starts off with a decent range in the morning and uses multiple short, fast charges during the day so that buses can complete their daily running of 200-225km per day. This kind of charging station needs a small space of 1 meter x 1 meter for installation and can be installed on the route and convenient locations. The objective should be to have more buses with higher frequencies so that passenger convenience is maximised.

Another method of charging is 'Battery Swapping' in which the bus stops at a charging station and gets a new battery and moves on while the old battery is taken out and charged. Battery swapping requires space to warehouse batteries, plus investment in automation to handle heavy batteries.

Battery Electric Bus

One of the most popular types of electric buses nowadays are battery electric buses. Battery electric buses have the electricity stored on board the vehicle in a battery. Today such buses can have a range of over 200 km with just one charge. These buses are usually used as city buses due to particularities in limited range. As a city bus, electric battery buses are cheaper than a comparable diesel buses.

City driving is majorly accelerating and braking. Due to this, the battery electric bus is superior to diesel bus as it can recharge most of the kinetic energy back into batteries in braking situations. This reduces brake wear on the buses and the use of electric over diesel can improve air quality in cities.

When operating within a city, it is important to minimize the unloaded and rolling weight of the bus. This can be accomplished by using aluminium as the main construction material for a bus. Composite paneling and other lightweight materials can also be used. According to Linkkebus their fully aluminium bus construction is about 3000 kg lighter than comparably-sized modern steel buses (curb weight 9500 kg). Reducing weight allows for a greater payload and reduces wear to components such as brakes, tires, and joints bringing cost savings to the operator annually.

Proterra's EcoRide BE35 transit bus, called the Ecoliner by Foothill Transit based in West Covina, California, is the world's first heavy duty, fast charge, battery-electric bus. Proterra's ProDrive drive-system uses a UQM motor and regenerative braking that captures 90% of the available energy and returns it to the TerraVolt energy storage system, which in turn increases the total distance the bus can drive by 31-35%. It can travel 30–40 miles on a single charge, is up to 600% more fuel-efficient than a typical diesel or CNG bus, and produces 44% less carbon than CNG.

Charging

Charging electric bus batteries is not as simple as refueling diesel engine. Special attention, monitoring, and scheduling are required to make optimal use of the charging process, while also ensuring proper battery maintenance and safekeeping. Some operators manage these challenges by purchasing extra buses. This way the charging can take place only at night. It is a safe solution, but also very costly and not scalable. The real solution is ensuring that the vehicle daily schedule takes into account also the need to charge, keeping the overall schedule as close to optimal as possible.

Today, there are various software companies that help bus operators manage their electric bus charging schedule. These solutions ensure that buses continue to operate safely, without any unplanned stops and inconvenience to passengers.

For communication between charger and electric bus the same ISO 15118 protocol is used as for passenger car charging. The only differences are in the charging power, voltage and coupler.

Pantographs and Underbody Collectors at Bus Stops

The lowered charging plate on an Arriva Shires & Essex Wright StreetLite EV bus whilst using induction to recharge its batteries at the Wolverton Agora bus stop.

Pantographs and underbody collectors are integrated in bus stops to quicken electric bus recharge, making it possible to use a smaller battery on the bus, which reduces the initial investment and subsequent costs.

Autonomous (Self-driving) Electric Buses

An autonomous bus is an electrically-powered, self-driving vehicle that transports twelve or more passengers. Autonomous buses are operated without a driver inside the vehicle, instead utilizing cameras, sensors and remote controls to properly steer their way through traffic.

Zinc-air Battery

There is a 40-foot (12.2 m) pure electric bus being developed, using a pre-commercial battery technology. Electric Fuel Corporation is developing and demonstrating a 40-foot (12.2 m) electric bus powered by a zinc air cell, along with an ultracapacitor. The zinc-air energy device, often described as a battery, converts zinc to zinc oxide in a process that provides energy to the bus. The bus is not recharged; instead, the zinc oxide cartridges are swapped out for new zinc ones. This bus has shown a range of over 100 miles (160 km) in testing and has been demonstrated in Las Vegas, Nevada. However, this technology is in the development phase, and several major hurdles must be overcome before it can be adopted for transit fleet use, including available refueling infrastructure or use in bus stations.

Capacitors Bus

Buses can use capacitors instead of batteries to store their energy. Ultracapacitors can only store about 5 percent of the energy that lithium-ion batteries hold for the same weight, limiting them to a couple of miles per charge. However ultracapacitors can charge and discharge much more rapidly than conventional batteries. In vehicles that have to stop frequently and predictably as part of normal operation, energy storage based exclusively on ultracapacitors can be a solution.

China is experimenting with a new form of electric bus, known as *Capabus*, which runs without continuous overhead lines by using power stored in large on-board electric double-layer capacitors, which are quickly recharged whenever the vehicle stops at any bus stop (under so-called electric umbrellas), and fully charged in the terminus.

A few prototypes were being tested in Shanghai in early 2005. In 2006, two commercial bus routes began to use electric double-layer capacitor buses; one of them is route 11 in Shanghai. In 2009, Sinautec Automobile Technologies, based in Arlington, VA, and its Chinese partner, Shanghai Aowei Technology Development Company are testing with 17 forty-one seat Ultracap Buses serving the Greater Shanghai area since 2006 without any major technical problems. Another 60 buses will be delivered early next year with ultracapacitors that supply 10 watt-hours per kilogram.

The buses have very predictable routes and need to stop regularly, every 3 miles (4.8 km), allowing opportunities for quick recharging. The trick is to turn some bus stops along the route into charging stations. At these stations, a collector on the top of the bus rises a few feet and touches an overhead charging line. Within a couple of minutes, the ultracapacitor banks stored under the bus seats are fully charged. The buses can also capture energy from braking, and the company says that recharging stations can be equipped with solar panels. A third generation of the product, will give 20 miles (32 km) of range per charge or better. Such a bus was delivered in Sofia, Bulgaria in May 2014 for 9 months' test. It covers 23 km in 2 charges.

Sinautec estimates that one of its buses has one-tenth the energy cost of a diesel bus and can achieve lifetime fuel savings of $200,000. Also, the buses use 40 percent less electricity compared to an electric trolley bus, mainly because they are lighter and have the regenerative braking benefits. The ultracapacitors are made of activated carbon, and have an energy density of six watt-hours per kilogram (for comparison, a high-performance lithium-ion battery can achieve 200 watt-hours per kilogram), but the ultracapacitor bus is also cheaper than lithium-ion battery buses, about 40 percent less expensive, with a far superior reliability rating.

There is also a plug-in hybrid version, which also uses ultracaps.

Future Developments

Sinautec is in discussions with MIT's Schindall about developing ultracapacitors of higher energy density using vertically aligned carbon nanotube structures that give the devices more surface area for holding a charge. So far, they are able to get twice the energy density of an existing ultracapacitor, but they are trying to get about five times. This would create an ultracapacitor with one-quarter of the energy density of a lithium-ion battery.

Future developments includes the use of inductive charging under the street, to avoid overhead wiring. A pad under each bus stop and at each stop light along the way would be used.

Surprising Facts about Electric Buses

Making the switch to sustainable transportation including zero emission electric buses will protect your passengers from the negative effects of harmful pollutants. Here are some surprising facts about electric buses that will lead you to reconsider the tangible benefits of choosing green power:

- Emissions: The impact of of emissions on children's health is undeniable. Diesel school buses emit NOx and PM 2.5 that cause asthma, which is affects 1 in 10 childern in the United states.
- Reflects Well on Your Brand: Individuals and corporations want to feel like they are contributing to a better world. As we become environmentally more conscious, customers are more inclined to perceive companies and institutions that align themselves with a commitment to sustainability in a more positive light than those that don't.
- Freedom from Rising Gasoline Costs: Gasoline fuel prices are often determined by geo-political and environmental factors. For instance, just two weeks after Hurricane Harvey hit Texas this year, several refineries closed, causing gas prices to spike. Using more electric transportation protects from the risk of global oil price hikes.
- Save on Maintenance Costs: Electric vehicles are composed of fewer moving parts compared to traditional internal combustion motors. Thus, electric buses cause fewer operational and maintenance issues.
- Quieter Operation: Though it may be difficult to imagine a quiet bus, electric buses do run silently and are therefore more neighborhood friendly.
- Reduced Costs: While electric buses for sale require a higher investment upfront, electricity is significantly cheaper then gasoline and can be partially funded with state and federal incentives. Electric buses also have fewer parts than standard buses. As a result, they have lower maintenance costs over time- another factor contributing to savings.

Advantages of the electric bus are:

- The cost of maintenance is about 25% lower than the maintenance cost of a diesel bus, since the electric motor doesn't need the same level of services the diesel engine needs.
- Electric engine's energy losses are significantly lower than diesel engine's energy losses, so the cost per kilometer of electric bus travel is about a third of the cost of a diesel bus ride.
- The electric bus is much quieter than a diesel bus, giving travelers a more pleasant experience and significantly reducing the environmental noise.
- The electric buses combine a regenerative braking system that generates electricity during the braking process.

Beside the positive side, there are also challenges:

- As of today, the electric bus is significantly more expensive than a diesel bus, but the cost is expected to go down.
- Adjustments must be made to the energy system according to the area in which the bus operates.
- The conventional charging system limits the driving distance. Charging systems that enable long driving distance are stile very expensive.
- Building a working plan (schedule) for an electric bus should consider a variety of parameters about the energy system.
- Real-time energy system monitoring is required to prevent malfunctions and delays.

References

- Interdisciplinary Design: Proceedings of the 21st CIRP Design Conference. Mary Kathryn Thompson. ISBN 9788989693291
- Advantages-and-disadvantages-of-owning-an-e-bike: geargadgetsandgizmos.com, Retrieved 12 April 2018
- Lee, S.; Huh, J.; Park, C.; Choi, N. S.; Cho, G. H.; Rim, C. T. (2010-09-01). "On-Line Electric Vehicle using inductive power transfer system". 2010 IEEE Energy Conversion Congress and Exposition: 1598–1601. doi:10.1109/ECCE.2010.5618092

- Electric-bus-demonstration, major-projects: winnipegtransit.com, Retrieved 22 April 2018
- "KAIST's Road-Embedded Recharger Named Among Best Inventions of 2010". The Chosun Ilbo. 2010-11-15. Retrieved 2010-11-15
- 5-benefit-of-electric-buses-from-creative-bus-sales: creativebussales.com, Retrieved 31 March 2018
- Spotnitz, R.; Franklin, J. (2003). "Abuse behavior of high-power, lithium-ion cells". Journal of Power Sources. 113: 81–100. doi:10.1016/S0378-7753(02)00488-3
- Advantages-challenges-electric-buses-gady-shlasky: linkedin.com, Retrieved 29 May 2018
- Sperling, Daniel and Deborah Gordon (2009). Two billion cars: driving toward sustainability. Oxford University Press, New York. pp. 22–26 and 114–139. ISBN 978-0-19-537664-7

2

Hybrid Electric Vehicles

Hybrid vehicles use more than one distinct power types. These can be two-wheeled and cycle-type vehicles or heavy vehicles. The aim of this chapter is to provide an understanding of hybrid electric vehicles, such as hybrid electric vehicle, plug-in hybrid electric vehicle, hybrid electric truck and hybrid electric bus, besides others.

Hybrid Vehicles

Hybrid vehicles are called hybrids because they use both a small internal combustion engine (ICE) and an electric motor to obtain maximum power and fuel economy with minimum emissions. How they do this varies from one model to another, with varying success.

What all hybrids have in common is the ability to generate electric current, store it in a large battery, and use that current to help drive the car. Hybrids capture electrical energy produced by a regenerative braking system, and their engines can power a generator, too. Hybrids can also conserve energy by shutting down the ICE when the vehicle is in Park, idling at a light, or stopped in traffic, or when the electric motor's energy is sufficient to drive the vehicle without assistance from the ICE.

Hybrids have regenerative braking systems that generate electric power to help keep the batteries charged. When the driver applies the brakes, the electric motor turns into a generator, and the magnetic drag slows the vehicle down. For safety, however, there is also a normal hydraulic braking system that can stop the car when regenerative braking isn't sufficient. There's no difference in maintenance or repair except that the brake pads tend to last much longer because they don't get used as much. In fact, if you drive a hybrid in a moderate manner, you almost never actually use the disc brakes on the wheels and may be able to go the life of the car without changing pads. The big difference is that regenerative brakes capture energy and turn it into electricity to charge the battery that provides power to an electric motor.

Parallel Hybrids

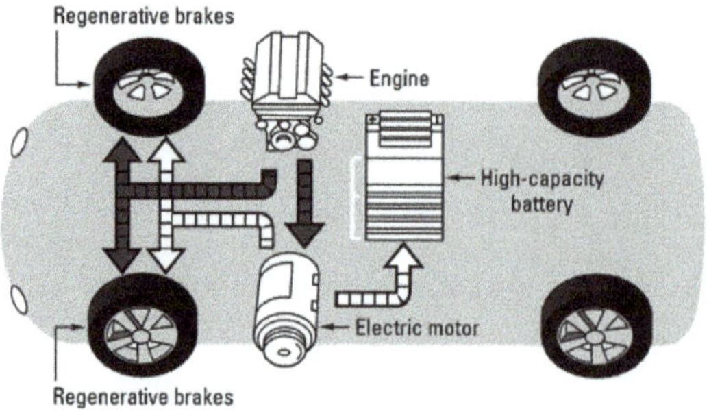

How a parallel hybrid works.

A parallel hybrid uses both an electric motor and an ICE for propulsion. They can run in tandem, or one can be used as the primary power source with the other kicking in to assist when extra power is needed for starting off, climbing hills, and accelerating to pass other vehicles. Because both are connected to the drive train, they're said to run "in parallel."

Series Hybrids

A series hybrid uses a gasoline or diesel ICE, coupled with a generator, to generate electricity but not to drive the car. The engine can send the electric current directly to the electric motor or charge a large battery that stores the electricity and delivers it to an electric motor on-demand. The electric motor propels the vehicle, using its power to rotate a driveshaft or a set of drive axles that turn the wheels.

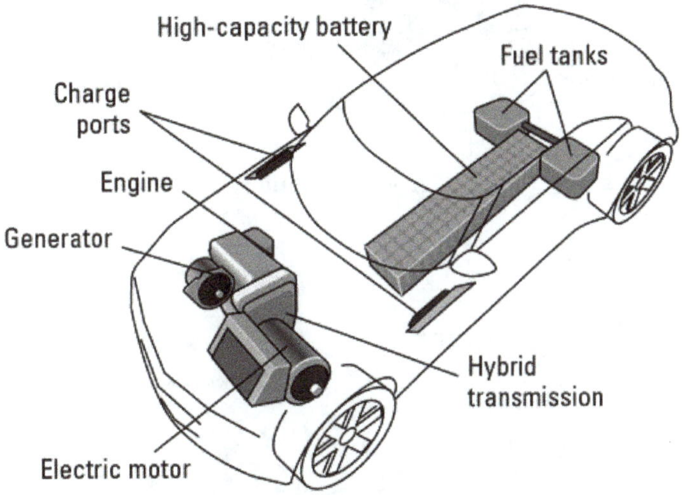

A series hybrid.

Plug-in Hybrids

Because plug-in hybrids feature larger batteries that can be charged at any ordinary 110-volt electrical socket, they have the capacity to extend the ability of the electric motor to drive the car farther without the need for starting the ICE and therefore substantially increase the vehicle's fuel efficiency. Estimates have ranged as high as 100 mpg.

Some technologically savvy individuals have adapted their hybrid vehicles into plug-in hybrids, and automakers are in the process of developing and producing them (sometimes in cooperation with major utility companies). The development of new, smaller, high-capacity lithium-ion batteries that can be recharged many times is the key to making plug-in hybrids available to the general public. Estimates are that plug-in hybrids equipped with these more powerful batteries will have a range of as much as 125 miles before the charge is depleted and the vehicle reverts to standard hybrid mode.

The main environmental problem with plug-in hybrids is that the electric current they draw is usually generated by utility companies powered by fossil fuels. The good news is that some major chains have committed to establishing charging stations powered by solar panels or wind energy, and many hybrid owners are willing to install solar panels to recharge these vehicles at home.

Plug-in hybrids charged by commercial sources of electricity or solar panels will be less dependent on the ICE, but will still need it for long trips, climbing hills, and so on. Future hybrids may use a small fuel cell to make electricity from hydrogen, which would mean the ICE would have to run even less frequently.

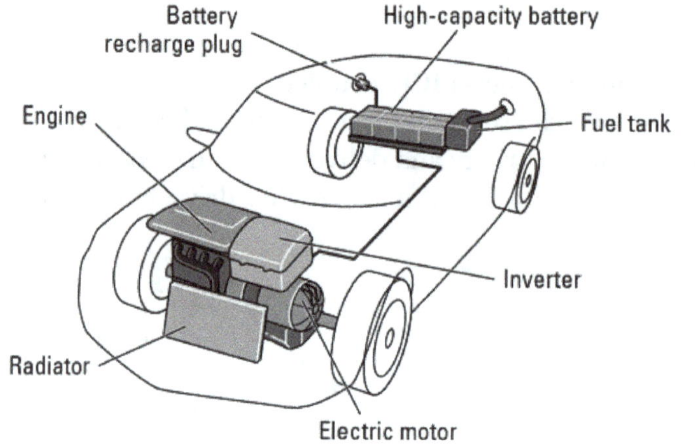

A plug-in hybrid

Two-mode Hybrids

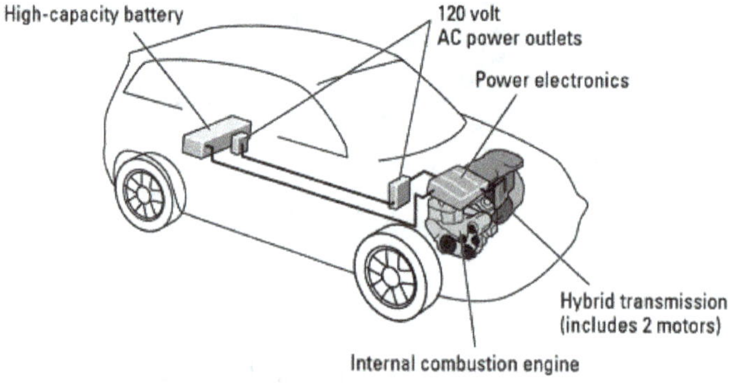

Two-mode hybrids may be the key to a competitive place for the U.S. in the hybrid market. Instead of the large storage battery found on conventional hybrids, two-mode hybrids use smaller batteries and two electric motors located inside an automatic transmission with two sets of gears — one for the ICE and the other to amplify the power of the electric motors. The transmission can function as a continuously variable transmission, as well. In one mode, at lower speeds, the vehicle can run with one or both electric motors, with or without the ICE, or on the ICE alone. At higher speeds, the second mode kicks in, and the ICE runs continuously in its higher gear ratios.

A two-mode Hybrid

Most hybrids use several advanced technologies:

- Regenerative Braking. Regenerative braking recaptures energy normally lost during coasting or braking. It uses the forward motion of the wheels to turn the motor. This generates electricity and helps slow the vehicle.

- Electric Motor Drive/Assist. The electric motor provides power to assist the engine in accelerating, passing, or hill climbing. This allows a smaller, more-efficient engine to be used. In some hybrids, the electric motor alone propels the vehicle at low speeds, where gasoline engines are least efficient.
- Automatic Start/Stop. Automatically shuts off the engine when the vehicle comes to a stop and restarts it when the accelerator is pressed. This reduces wasted energy from idling.

Hybrid Car

Hybrid cars are becoming more popular and more common. Basically, a hybrid car is one that uses two or more engines i.e. an electric motor and a conventional engine (either petrol or diesel). The electric engine powers the car at lower speeds and gas engine powers it at higher speeds. A hybrid car like Toyota Prius and Civic Hybrid not only conserves fuel but also produce less CO_2 emissions. Though hybrid vehicles are now growing in popularity but still few people are actually using it mainly due to lack of knowledge of how hybrid vehicles work and whether they're as good as other gasoline powered vehicles.

While the technology has existed since the early 1900's, it has only been in the past decade or so that the price of manufacturing them has brought them into the range of possibility for the average driver. There are also more government incentive programs that use credits and special discounts to support the purchase and use of hybrid vehicles. Many cities are switching their public transportation and service vehicles over to hybrid cars and buses as a part of the program to become more environmentally responsible.

Advantages of a Hybrid Car

Here are few of the top advantages of having a hybrid car:

- Environmentally Friendly: One of the biggest advantage of hybrid car over gasoline powered car is that it runs cleaner and has better gas mileage which makes it environmentally friendly. A hybrid vehicle runs on twin powered engine (gasoline engine and electric motor) that cuts fuel consumption and conserves energy.
- Financial Benefits: Hybrid cars are supported by many credits and incentives that help to make them affordable. Lower annual tax bills and exemption from congestion charges comes in the form of less amount of money spent on the fuel.
- Less dependence on Fossil Fuels: A Hybrid car is much cleaner and requires less fuel to run which means less emissions and less dependence on fossil fuels. This in turn also helps to reduce the price of gasoline in domestic market.
- Regenerative Braking System: Each time you apply brake while driving a hybrid vehicle helps you to recharge your battery a little. An internal mechanism kicks in that captures the energy released and uses it to charge the battery which in turn eliminates the amount of time and need for stopping to recharge the battery periodically.
- Built From Light Materials: Hybrid vehicles are made up of lighter materials which means less energy is required to run. The engine is also smaller and lighter which also saves much energy.

- Higher Resale Value: With continuous increase in price of gasoline, more and more people are turning towards hybrid cars. The result is that these green vehicles have started commanding higher than average resale values. So, in case you are not satisfied with your vehicle, you can always sell it at a premium price to buyers looking for it.

There are many advantages to owning a hybrid car. The one you will like the best is how it helps you to control your budget as gas prices continue to get higher. The other benefit that is not seen directly is how owning and driving a hybrid car impacts the environment. It reduces the dependence on fossil fuels and lowers your carbon imprint on the environment.

Disadvantages of a Hybrid Car

There are disadvantages to owning a hybrid car, but they are probably not what you think. Contrary to popular myth, hybrid cars have just as much power as regular cars and have no issue with mountain driving or towing. The disadvantages will depend on the type of hybrid fuel that your car uses.

Here are few of the disadvantages of a hybrid car:

- Less Power: Hybrid cars are twin powered engine. The gasoline engine which is primary source of power is much smaller as compared to what you get in single engine powered car and electric motor is low power. The combined power of both is often less than that of gas powered engine. It is therefore suited for city driving and not for speed and acceleration.
- Can be Expensive: The biggest drawback of having a hybrid car is that it can burn a hole in your pocket. Hybrid cars are comparatively expensive than a regular petrol car and can cost $5000 to $10000 more than a standard version. However, that extra amount can be offset with lower running cost and tax exemptions.
- Poorer Handling: A hybrid car houses an gasoline powered engine, a lighter electric engine and a pack of powerful batteries. This adds weight and eats up the extra space in the car. Extra weight results in fuel inefficiency and manufacturers cut down weight which has resulted in motor and battery downsizing and less support in the suspension and body.
- Higher Maintenance Costs: The presence of dual engine, continuous improvement in technology, and higher maintenance cost can make it difficult for mechanics to repair the car. It is also difficult to find a mechanic with such an expertise.
- Presence of High Voltage in Batteries: In case of an accident, the high voltage present inside the batteries can prove lethal for you. There is a high chance of you getting electrocuted in such cases which can also make the task difficult for rescuers to get other passengers and driver out of the car.

Hybrid Vehicle Drivetrain

A drivetrain is the collection of components that deliver power from a vehicle's engine or motor to the vehicle's wheels. In hybrid-electric cars, the drivetrain's design determines how the electric motor works in conjunction with the conventional engine. The drivetrain affects the vehicle's mechanical efficiency, fuel consumption, and purchasing price.

Hybrids that use a series drivetrain only receive mechanical power from the electric motor, which is run by either a battery or a gasoline-powered generator. In hybrids with parallel drivetrains, the electric motor and internal combustion engine can provide mechanical power simultaneously. Series/parallel drivetrains enable the engine and electric motor to provide power independently or in conjunction with one another.

Both conventional hybrids and plug-in hybrids have models with series, parallel, and series/parallel drivetrains. Since battery-electric and hydrogen fuel cell vehicles don't have internal combustion engines, they utilize different drivetrain assemblies (though some components are shared).

Series Hybrid

In a series hybrid system, the combustion engine drives an electric generator (usually a three-phase alternator plus rectifier) instead of directly driving the wheels. The electric motor is the only means of providing power to the wheels. The generator both charges a battery and powers an electric motor that moves the vehicle. When large amounts of power are required, the motor draws electricity from both the batteries and the generator.

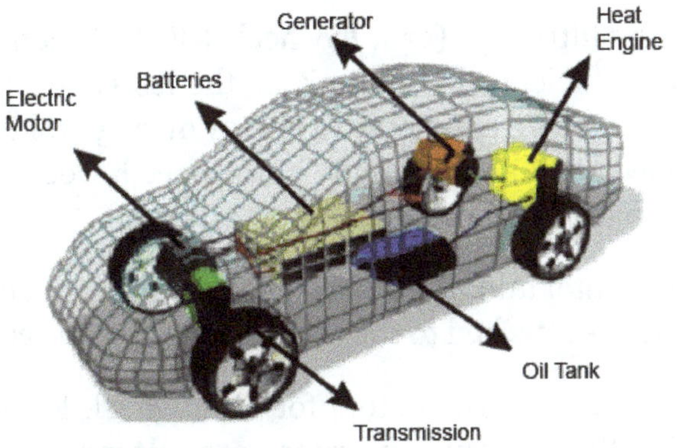

Series hybrid configurations already exist a long time: diesel-electric locomotives, hydraulic earth moving machines, diesel-electric power groups, loaders.

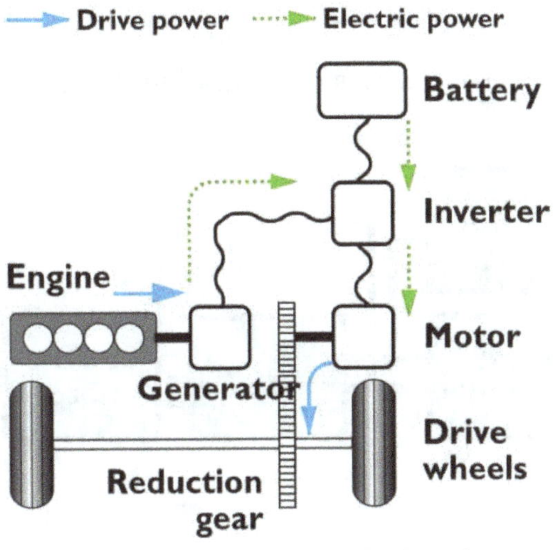

Structure of a series hybrid vehicle (below with flywheel or ultracaps as peak power unit)

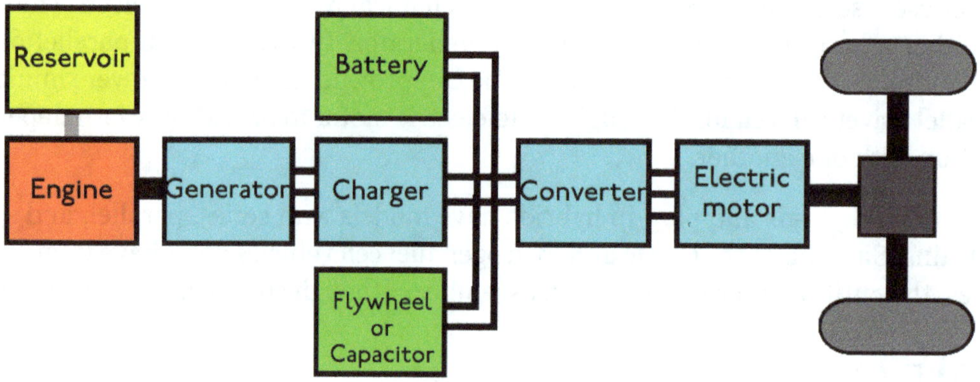

Series hybrids can be assisted by ultracaps (or a flywheel: KERS=Kinetic Energy Recuperation System), which can improve the efficiency by minimizing the losses in the battery. They deliver peak energy during acceleration and take regenerative energy during braking. Therefore, the ultracaps are kept charged at low speed and almost empty at top speed. Deep cycling of the battery is reduced, the stress factor of the battery is lowered.

Series hybrids can be assisted by ultracaps (or a flywheel: KERS=Kinetic Energy Recuperation System), which can improve the efficiency by minimizing the losses in the battery. They deliver peak energy during acceleration and take regenerative energy during braking. Therefore, the ultracaps are kept charged at low speed and almost empty at top speed. Deep cycling of the battery is reduced, the stress factor of the battery is lowered.

A complex transmission between motor and wheel is not needed, as electric motors are efficient over a wide speed range. If the motors are attached to the vehicle body, flexible couplings are required.

Some vehicle designs have separate electric motors for each wheel. Motor integration into the wheels has the disadvantage that the unsprung mass increases, decreasing ride performance. Advantages of individual wheel motors include simplified traction control (no conventional mechanical transmission elements such as gearbox, transmission shafts, differential), all wheel drive, and allowing lower floors, which is useful for buses. Some 8x8 all-wheel drive military vehicles use individual wheel motors.

A fuel cell hybrid electric always has a series configuration: the engine-generator combination is replaced by a fuel cell.

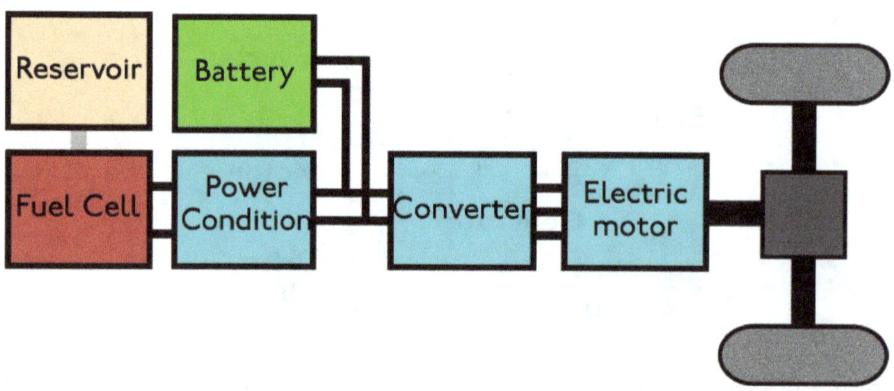

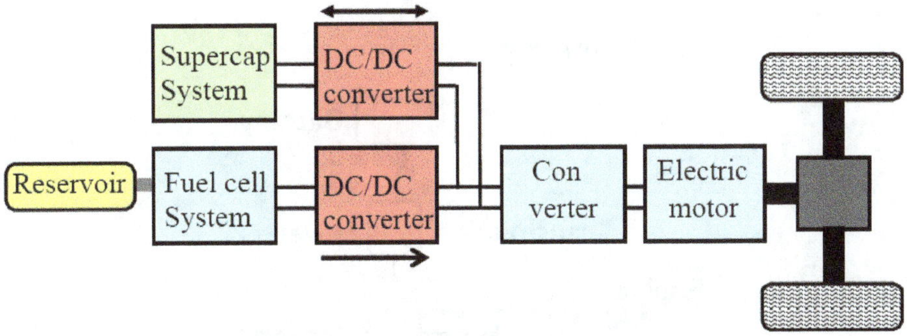

Structures of a fuel cell hybrid electric vehicle

Weaknesses of series hybrid vehicles:

- The ICE, the generator and the electric motor are dimensioned to handle the full power of the vehicle. Therefore, the total weight, cost and size of the powertrain can be excessive.
- The power from the combustion engine has to run through both the generator and electric motor. During long-distance highway driving, the total efficiency is inferior to a conventional transmission, due to the several energy conversions.

Advantages of series hybrid vehicles:

- There is no mechanical link between the combustion engine and the wheels. The engine-generator group can be located everywhere.
- There are no conventional mechanical transmission elements (gearbox, transmission shafts).
- Separate electric wheel motors can be implemented easily.
- The combustion engine can operate in a narrow rpm range (its most efficient range), even as the car changes speed.
- Series hybrids are relatively the most efficient during stop-and-go city driving.

Parallel Hybrid

Parallel hybrid systems have both an internal combustion engine (ICE) and an electric motor in parallel connected to a mechanical transmission.

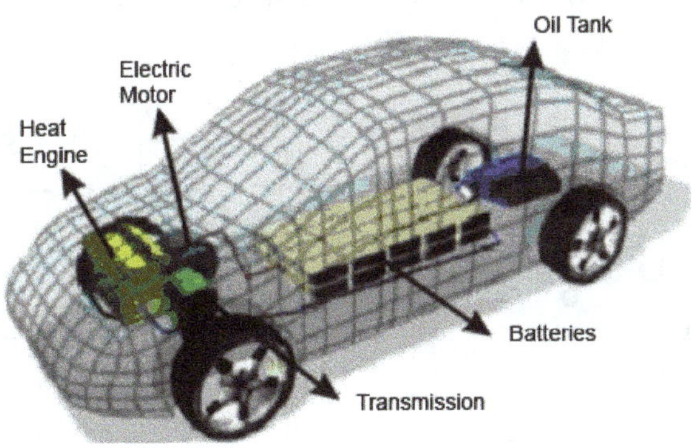

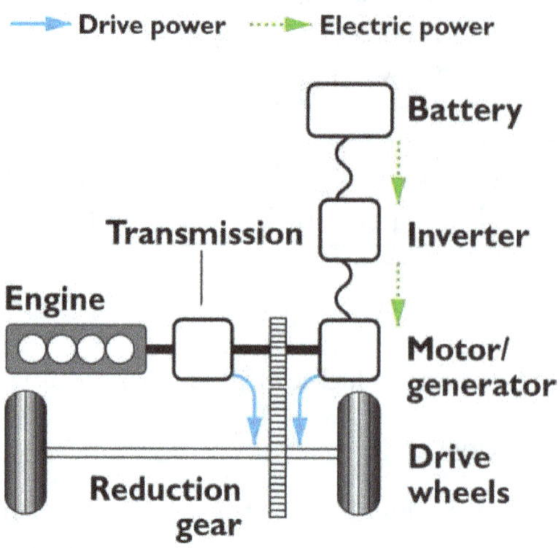

Structure of a parallel hybrid electric vehicle

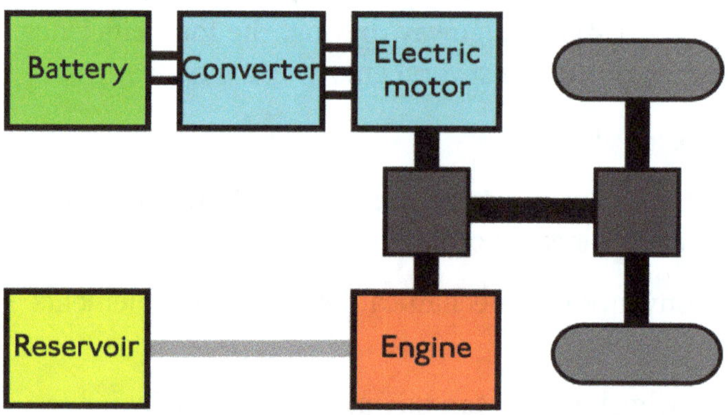

Most designs combine a large electrical generator and a motor into one unit, often located between the combustion engine and the transmission, replacing both the conventional starter motor and the alternator. The battery can be recharged during regenerative breaking, and during cruising (when the ICE power is higher than the required power for propulsion). As there is a fixed mechanical link between the wheels and the motor (no clutch), the battery cannot be charged when the car isn't moving.

When the vehicle is using electrical traction power only, or during brake while regenerating energy, the ICE is not running (it is disconnected by a clutch) or is not powered (it rotates in an idling manner).

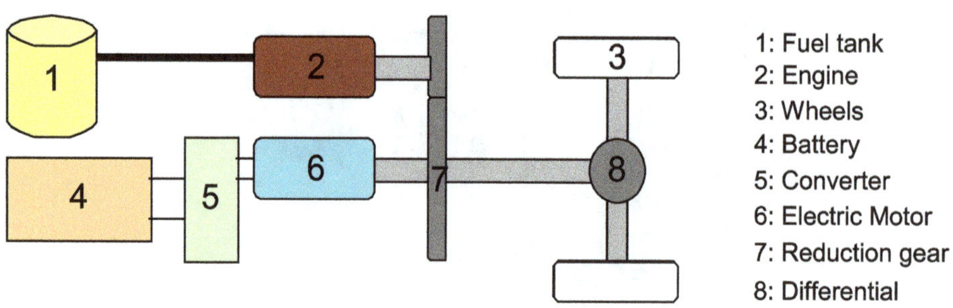

$$\omega_8 = \omega_6 = \omega_2 / x_2$$

$$T_8 = \frac{T_8}{\eta_6} + \frac{x_2 . T_2}{\eta_2}$$

Operation Modes:

The parallel configuration supports diverse operating modes:

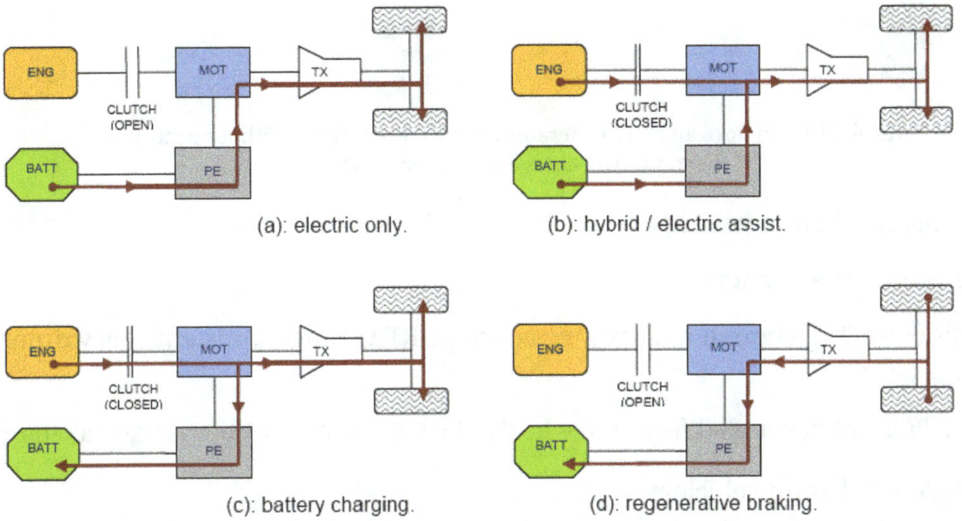

Some typical modes for a parallel hybrid configuration PE = Power electronics
TX = Transmission

- Electric power only: Up to speeds of usually 40 km/h, the electric motor works with only the energy of the batteries, which are not recharged by the ICE. This is the usual way of operating around the city, as well as in reverse gear, since during reverse gear the speed is limited.
- ICE power only: At speeds superior to 40 km/h, only the heat engine operates. This is the normal operating way at the road.
- ICE + electric power: if more energy is needed (during acceleration or at high speed), the electric motor starts working in parallel to the heat engine, achieving greater power
- ICE + battery charging: if less power is required, excess of energy is used to charge the batteries. Operating the engine at higher torque than necessary, it runs at a higher efficiency.
- Regenerative breaking: While braking or decelerating, the electric motor takes profit of the kinetic energy of the he moving vehicle to act as a generator.
- Sometimes, an extra generator is used: then the batteries can be recharged when the vehicle is not driving, the ICE operates disconnected from the transmission. But this system gives an increased weight and price to the HEV.

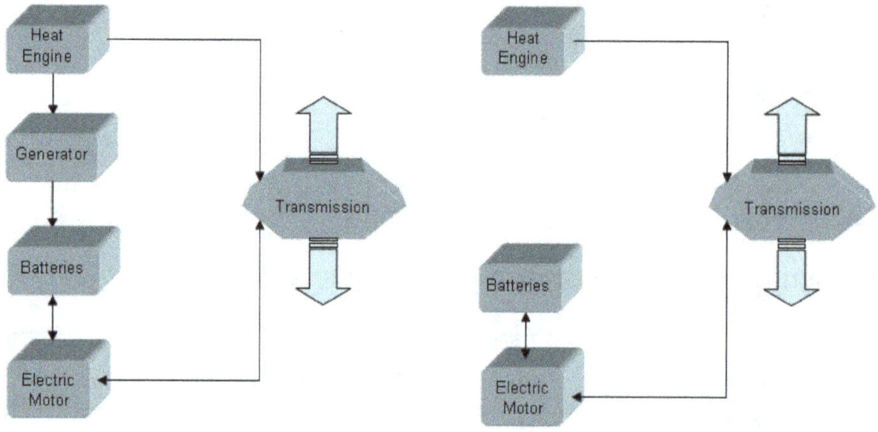

A parallel HEV can have an extra generator for the battery (left) Without generator, the motor will charge the battery (right)

Weaknesses of parallel hybrid vehicles:

- Rather complicated system.
- The ICE doesn't operate in a narrow or constant RPM range, thus efficiency drops at low rotation speed.
- As the ICE is not decoupled from the wheels, the battery cannot be charged at standstill.

Advantages of parallel hybrid vehicles:

- Total efficiency is higher during cruising and long-distance highway driving.
- Large flexibility to switch between electric and ICE power
- Compared to series hybrids, the electromotor can be designed less powerful than the ICE, as it is assisting traction. Only one electrical motor/generator is required.

Example of PHEV:

Honda Civic: Honda's IMA (Integrated Motor Assist) uses a rather traditional ICE with continuously variable transmission, where the flywheel is replaced with an electric motor.

Influence of scale: a Volvo 26 ton truck (12 ton own weight, 14 ton max load) equipped with 200 kg of batteries can drive on pure electric power for 2 minutes only! Because of space constraints, it is not possible to build in more batteries.

BMW 7Series ActiveHybrid.

Combined Hybrid

Combined hybrid systems have features of both series and parallel hybrids. There is a double connection between the engine and the drive axle: mechanical and electrical. This split power path allows interconnecting mechanical and electrical power, at some cost in complexity.

Power-split devices are incorporated in the powertrain. The power to the wheels can be either mechanical or electrical or both. This is also the case in parallel hybrids. But the main principle behind the combined system is the decoupling of the power supplied by the engine from the power demanded by the driver.

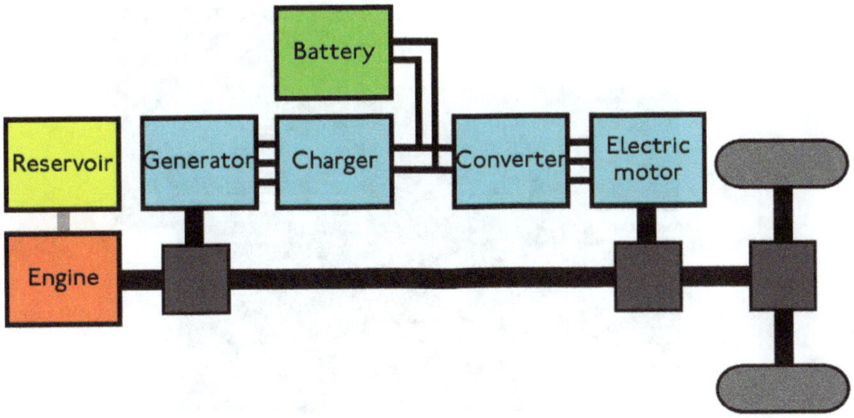

Simplified structure of a combined hybrid electric vehicle

In a conventional vehicle, a larger engine is used to provide acceleration from standstill than one needed for steady speed cruising. This is because a combustion engine's torque is minimal at lower RPMs, as the engine is its own air pump. On the other hand, an electric motor exhibits maximum torque at stall and is well suited to complement the engine's torque deficiency at low RPMs. In a combined hybrid, a smaller, less flexible, and highly efficient engine can be used. It is often a variation of the conventional Otto cycle, such as the Miller or Atkinson cycle. This contributes significantly to the higher overall efficiency of the vehicle, with regenerative braking playing a much smaller role.

At lower speeds, this system operates as a series HEV, while at high speeds, where the series powertrain is less efficient, the engine takes over. This system is more expensive than a pure parallel system as it needs an extra generator, a mechanical split power system and more computing power to control the dual system.

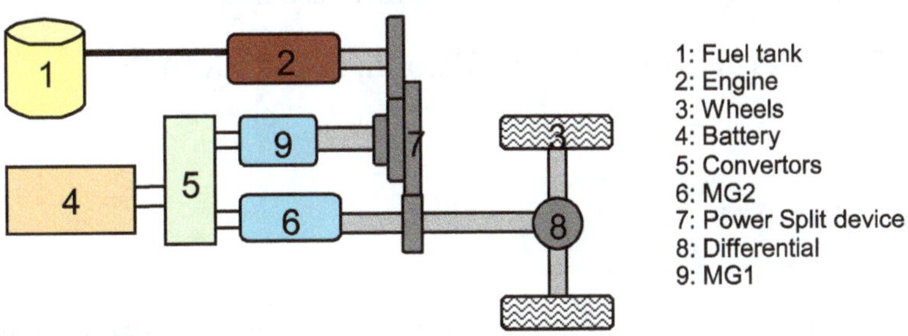

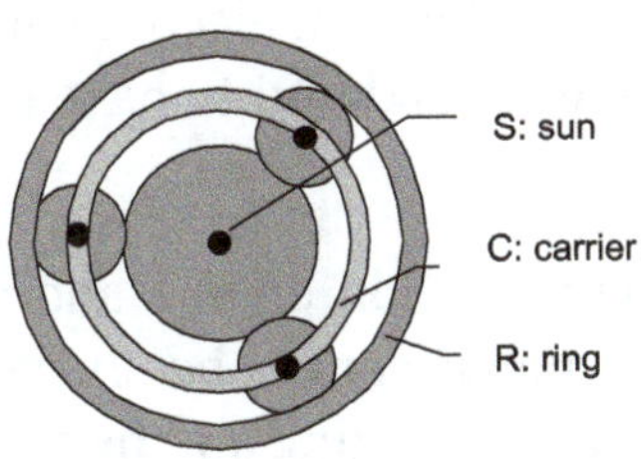

$$(1+\rho).\ \omega_c = \omega_s + \rho.\omega_R$$

$$\rho = z_R / z_S$$

Combined HEV with planetary unit as used in the Toyota Prius

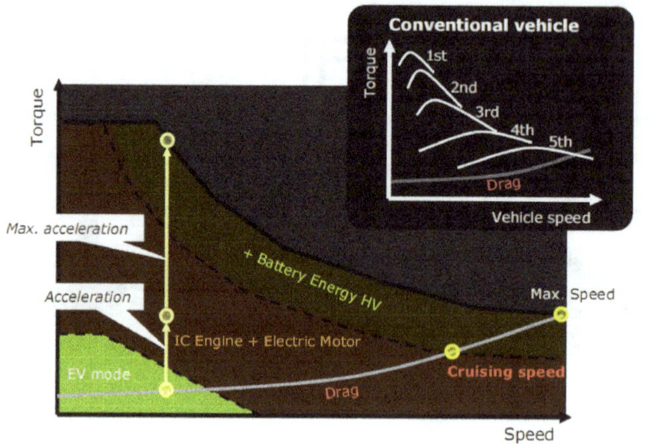

Combined hybrid drive modes

Weaknesses of combined hybrid vehicles:

- Very complicated system, more expensive than parallel hybrid.
- The efficiency of the power train transmission is dependent on the amount of power being transmitted over the electrical path, as multiple conversions, each with their own efficiency, lead to a lower efficiency of that path (~70%) compared with the purely mechanical path (98%).

Advantages of combined hybrid vehicles:

- Maximum flexibility to switch between electric and ICE power.
- Decoupling of the power supplied by the engine from the power demanded by the driver allows for a smaller, lighter, and more efficient ICE design.

Example of CHEV: Toyota Prius, Auris, Lexus CT200h, Lexus RX400h.

Hybrid Electric Vehicle

A hybrid electric vehicle (HEV) is a type of vehicle that uses both an electric engine and a conventional internal combustion engine. This type of vehicle is considered to have better performance and fuel economy compared to a conventional one.

Hybrid electric vehicles are powered by an internal combustion engine and an electric motor, which uses energy stored in batteries. A hybrid electric vehicle cannot be plugged in to charge the battery. Instead, the battery is charged through regenerative braking and by the internal combustion engine. The extra power provided by the electric motor can potentially allow for a smaller engine. The battery can also power auxiliary loads like sound systems and headlights, and reduce engine idling when stopped. Together, these features result in better fuel economy without sacrificing performance. Learn more about hybrid electric vehicles.

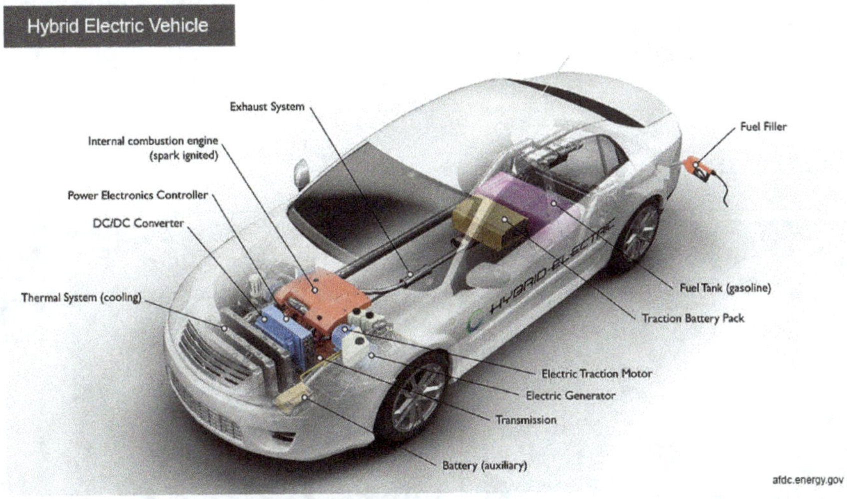

Key Components of a Hybrid Electric Car

Battery (auxiliary): In an electric drive vehicle, the auxiliary battery provides electricity to start the car before the traction battery is engaged and also powers vehicle accessories.

DC/DC converter: This device converts higher-voltage DC power from the traction battery pack to the lower-voltage DC power needed to run vehicle accessories and recharge the auxiliary battery.

Electric generator: Generates electricity from the rotating wheels while braking, transferring that energy back to the traction battery pack. Some vehicles use motor generators that perform both the drive and regeneration functions.

Electric traction motor: Using power from the traction battery pack, this motor drives the vehicle's wheels. Some vehicles use motor generators that perform both the drive and regeneration functions.

Exhaust system: The exhaust system channels the exhaust gases from the engine out through the tailpipe.

Fuel filler: This is a filler or "nozzle" used to add fuel to the tank.

Fuel tank (gasoline): This tank stores gasoline on board the vehicle until it's needed by the engine.

Internal combustion engine (spark-ignited): In this configuration, fuel is injected into either the intake manifold or the combustion chamber, where it is combined with air, and the air/fuel mix is ignited by the spark from a spark plug.

Power electronics controller: This unit manages the flow of electrical energy delivered by the traction battery, controlling the speed of the electric traction motor and the torque it produces.

Thermal system (cooling): This system maintains a proper operating temperature range of the engine, electric motor, power electronics, and other components.

Traction battery pack: Stores electricity for use by the electric traction motor.

Transmission: The transmission transfers mechanical power from the engine and/or electric traction motor to drive the wheels.

Classification

Types of Powertrain

The Toyota Highlander Hybrid has a series-parallel drivetrain.

The Saturn Vue Green Line is a mild hybrid.

The BMW Concept 7 Series ActiveHybrid is a mild hybrid with an electric motor designed to increase power and performance.

The 2005–06 Chevrolet Silverado Hybrid is a mild hybrid using the electric motor mainly to power the accessories.

Hybrid electric vehicles can be classified according to the way in which power is supplied to the drivetrain:

- In parallel hybrids, the ICE and the electric motor are both connected to the mechanical transmission and can simultaneously transmit power to drive the wheels, usually through a conventional transmission. Honda's Integrated Motor Assist (IMA) system as found in the Insight, Civic, Accord, as well as the GM Belted Alternator/Starter (BAS Hybrid) system found in the Chevrolet Malibu hybrids are examples of production parallel hybrids. The internal combustion engine of many parallel hybrids can also act as a generator for supplemental recharging. As of 2013, commercialized parallel hybrids use a full size combustion engine with a single, small (<20 kW) electric motor and small battery pack as the electric motor is designed to supplement the main engine, not to be the sole source of motive power from launch. But after 2015 parallel hybrids with over 50 kW are available, enabling electric driving at moderate acceleration. Parallel hybrids are more efficient than comparable non-hybrid vehicles especially during urban stop-and-go conditions where the electric motor is permitted to contribute, and during highway operation.

- In series hybrids, only the electric motor drives the drivetrain, and a smaller ICE (also called range extender) works as a generator to power the electric motor or to recharge the batteries. They also usually have a larger battery pack than parallel hybrids, making them more expensive. Once the batteries are low, the small combustion engine can generate power at its optimum settings at all times, making them more efficient in extensive city driving.
- Power-split hybrids have the benefits of a combination of series and parallel characteristics. As a result, they are more efficient overall, because series hybrids tend to be more efficient at lower speeds and parallel tend to be more efficient at high speeds; however, the cost of power-split hybrid is higher than a pure parallel. Examples of power-split (referred to by some as "series-parallel") hybrid powertrains include 2007 models of Ford, General Motors, Lexus, Nissan, and Toyota.

In each of the hybrids above it is common to use regenerative braking to recharge the batteries.

Types by Degree of Hybridization

Full hybrid, sometimes also called a strong hybrid, is a vehicle that can run only on a combustion engine, only on an electric motor, or a combination of both. Ford's hybrid system, Toyota's Hybrid Synergy Drive and General Motors/Chrysler's Two-Mode Hybrid technologies are full hybrid systems. The Toyota Prius, Ford Escape Hybrid, and Ford Fusion Hybrid are examples of full hybrids, as these cars can be moved forward on battery power alone. A large, high-capacity battery pack is needed for battery-only operation. These vehicles have a split power path allowing greater flexibility in the drivetrain by interconverting mechanical and electrical power, at some cost in complexity.

Mild hybrid, is a vehicle that cannot be driven solely on its electric motor, because the electric motor does not have enough power to propel the vehicle on its own. Mild hybrids include only some of the features found in hybrid technology, and usually achieve limited fuel consumption savings, up to 15 percent in urban driving and 8 to 10 percent overall cycle. A mild hybrid is essentially a conventional vehicle with oversize starter motor, allowing the engine to be turned off whenever the car is coasting, braking, or stopped, yet restart quickly and cleanly. The motor is often mounted between the engine and transmission, taking the place of the torque converter, and is used to supply additional propulsion energy when accelerating. Accessories can continue to run on electrical power while the gasoline engine is off, and as in other hybrid designs, the motor is used for regenerative braking to recapture energy. As compared to full hybrids, mild hybrids have smaller batteries and a smaller, weaker motor/generator, which allows manufacturers to reduce cost and weight.

Honda's early hybrids including the first generation Insight used this design, leveraging their reputation for design of small, efficient gasoline engines; their system is dubbed Integrated Motor Assist (IMA). Starting with the 2006 Civic Hybrid, the IMA system now can propel the vehicle solely on electric power during medium speed cruising. Another example is the 2005–2007 Chevrolet Silverado Hybrid, a full-size pickup truck. Chevrolet was able to get a 10% improvement on the Silverado's fuel efficiency by shutting down and restarting the engine on demand and using regenerative braking. General Motors has also used its mild BAS Hybrid technology in other models such as the Saturn Vue Green Line, the Saturn Aura Greenline and the Malibu Hybrid.

Plug-in Hybrids (PHEVs)

The BYD F3DM is the world's first mass-produced plug-in hybrid with an all-electric range of 40 to 60 miles (64 to 96 km).

The Chevrolet Volt is a plug-in hybrid able to run in all-electric mode up to 35 miles (56 km).

A plug-in hybrid electric vehicle (PHEV), also known as a plug-in hybrid, is a hybrid electric vehicle with rechargeable batteries that can be restored to full charge by connecting a plug to an external electric powersource. A PHEV shares the characteristics of both a conventional hybrid electric vehicle, having an electric motor and an internal combustion engine; and of an all-electric vehicle, also having a plug to connect to the electrical grid. PHEVs have a much larger all-electric range as compared to conventional gasoline-electric hybrids, and also eliminate the "range anxiety" associated with all-electric vehicles, because the combustion engine works as a backup when the batteries are depleted.

Chinese battery manufacturer and automaker BYD Auto released the F3DM PHEV-62 (PHEV-100 km) hatchback to the Chinese fleet market on December 15, 2008, for 149,800 yuan (US $22,000). General Motors launched the 2011 Chevrolet Volt series plug-in in December 2010. At the time, the Volt displaced the Toyota Prius as the most fuel-efficient car sold in the United States.

As of December 2016, the Volt/Ampera family is the world's all-time best-selling plug-in hybrid car, with global sales totaling about 134,500 units since its inception, including over 10,000 Opel/

Vauxhall Amperas sold in Europe. The Mitsubishi Outlander P-HEV ranks second with about 119,500 units delivered worldwide. Third is the Toyota Prius Plug-in Hybrid with cumulative global sales of 79,300 units at the end of January 2017.

Sales and Rankings

As of January 2017, more than 12 million hybrid electric vehicles have been sold worldwide since their inception in 1997. As of April 2016, Japan ranked as the market leader with more than 5 million hybrids sold, followed by the United States with cumulative sales of over 4 million units since 1999, Europe with about 1.5 million, and the rest of the world with combined sales of over 500,000 hybrids. As of August 2014, more than 130,000 hybrids have been sold in Canada, of which, over 100,000 are Toyota and Lexus models. In Australia, over 50,000 Lexus and Toyota models have been sold through February 2014.

As of January 2017, global hybrid sales led by Toyota Motor Company (TMC) with over 10 million Lexus and Toyota hybrids sold; followed by Honda Motor Co., Ltd. with cumulative global sales of more than 1.35 million hybrids as of June 2014; Ford Motor Corporation with over 424,000 hybrids sold in the United States through June 2015, of which, around 10% are plug-in hybrids; Hyundai Group with cumulative global sales of 200,000 hybrids as of March 2014, including both Hyundai Motors and Kia Motors hybrid models; and PSA Peugeot Citroën with over 50,000 diesel-powered hybrids sold in Europe through December 2013.

The Toyota Prius is the all-time top-selling hybrid in both Japan and the U.S.

TMC experienced record sales of hybrid cars during 2013, with 1,279,400 units sold worldwide, and it took nine months to achieve one million hybrid sales. Again in 2014, TMC sold a record one million hybrids in nine months. Toyota hybrids combined with Lexus models reached 1 million units in May 2007, and the U.S. reached the 1 million mark of sales of both brands by February 2009. Worldwide sales of TMC hybrids totaled over 2 million vehicles by August 2009, 3 million units by February 2011, 5 million in March 2013, 7 million in September 2014, and the 8 million mark in July 2015. The 9 million sales mark was reached in April 2016, again, selling one million hybrids in just ninth months, and the 10 million milestone in January 2017, achieved one more time just nine months after the previous million.

Ford experienced record sales of its hybrids models in the U.S. during 2013, with almost 80,000 units sold, almost triple the 2012 total. During the second quarter of 2013 Ford achieved its best hybrid sales quarter ever, up 517% over the same quarter of 2012. In 2013 Toyota's hybrid market share in the U.S. declined from 2012 totals due to new competition, particularly from Ford with the arrival of new products such as the C-Max Hybrid and the new styling of the Fusion. Except for the Prius c, sales of the other models of the Prius family and the Camry Hybrid suffered a decline from 2012, while the Fusion Hybrid experienced a 164.3% increase from 2012, and C-Max Hybrid sales climbed 156.6%. During 2013 Ford increased its market share of the American hybrid market from 7.5% in 2012 to 14.7% in 2013.

As of January 2017, global hybrid sales are led by the Prius family, with cumulative sales of 6.0361 million units (excluding plug-in hybrids) representing 60% of the 10 million hybrids sold worldwide by Toyota and Lexus since 1997. As of January 2017, the Toyota Prius liftback is the leading model of the Toyota brand with cumulative sales of 3.985 million units. Ranking second is the Toyota Aqua/Prius c, with global sales of 1.380 million units, followed by the Prius v/α/+ with 671,200, the Camry Hybrid with 614,700 units, the Toyota Auris with 378,000 units, and the Toyota Yaris Hybrid with 302,700. U.S. sales of the Toyota Prius reached the 1.0 million milestone in early April 2011, and cumulative sales of the Prius in Japan exceeded the 1 million mark in August 2011. As of January 2017, sales of the Prius liftback totaled over 1.8 million units in Japan and 1.75 million in the United States, ranking as the all-time best-selling hybrid car in both countries.

Global sales of Lexus brand hybrid vehicles worldwide reached the 500 thousand mark in November 2012. The 1 million sales milestone was achieved in March 2016. The Lexus RX 400h/RX 450h ranks as the top-selling Lexus hybrid with 363,000 units delivered worldwide as of January 2017, followed by the Lexus CT 200h with 290,800 units, and the Lexus ES 300h with 143,200 units.

Annual Sales in Top Markets

Top national markets for hybrid electric vehicles between 2007 and 2015									
Country	Number of hybrids sold or registered by year								
	2015	2014	2013	2012	2011	2010	2009	2008	2007
Japan	633,200[1]	Over 1 million	679,100[1]	678,000[1]	316,300[1]	392,200[1]	334,000	94,259	69,015
US	384,404	452,152	495,771	434,498	268,752	274,210	290,271	312,386	352,274
France	56,030	41,208	46,785	27,730	13,340	9,443	9,399	9,137	7,268
UK	44,580	37,215	29,129	24,900	23,391	22,127	14,645	15,385	15,971
Italy	25,240	21,154	14,695	5,885	5,244				
Germany	22,529	22,908	24,963	21,438	12,622	10,661	8,374	6,464	7,591
Spain	18,406	12,083	10,294	10,030	10,350				
Netherlands	13,752	10,341	18,356	19,519	14,874	16,111	16,122	11,837	3,013
Canada	Not available		~15,000	~25,000	Not available		16,167	19,963	14,828
World	Over 1.2 million	Over 1.6 million	Over 1.3 million	Over 1.2 million	-	-	740,000	511,758	500,405
(1) Partial sales, includes only Toyota/Lexus sales. (2) French registrations between 2011 and 2013 include plug-in hybrids									

As of April 2016, Japan ranks as the world's market leader with more than 5 million hybrids sold since 1997, representing around 45% of cumulative global hybrid sales since their inception. After 18 years since their introduction in the Japanese market, annual hybrid sales surpassed the 1 million mark for the first time in 2014. With cumulative sales of over 4 million hybrids through December 2014, Japan surpassed the United States as the world's largest hybrid market. It was also the first time that all eight major Japanese manufacturers offered hybrid vehicles in their lineup.

Japan also has the world's highest hybrid market penetration. The hybrid market share of new car sales began to increase significantly in 2009, when the government implemented aggressive fiscal incentives for fuel efficient vehicles and the third generation Prius was introduced. That year, the hybrid market share of new car sales in the country, including kei cars, jumped from less than 5% in 2008 to over 10% in 2009. If only conventional passenger cars are accounted for, the hybrid market share was about 15%. By 2013 the hybrid market share accounted for more than 30% of the 2.9 million standard passenger vehicles sold, and about 20% of the 4.5 million passenger vehicles including kei cars. Sales of standard cars in 2016 totaled 1.49 million units, with the hybrid segment achieving a record 38% market share. Accounting for kei cars, hybrids achieved a market share of 25.7% of new passenger car sales, up from 22.3% in 2015. In 2016 every one of the standard cars listed in the Japanese top-20 best-selling car ranking had a hybrid version on sale. and the two top-selling standard cars were models available only as a hybrid, the Toyota Prius and the Toyota Aqua.

Toyota's hybrid sales in Japan since 1997, including both Toyota and Lexus models, passed the 1 million mark in July 2010, 2 million in October 2012, and topped the 3 million mark in March 2014. As of January 2017, TMC hybrid sales in the country totaled 4,853,000 vehicles, of which, only 4,900 units are commercial vehicles. Cumulative sales of the original Prius in Japan reached the 1 million mark in August 2011. Sales of the Prius family vehicles totaled 3,435,800 units through January 2017. The Prius liftback is the top-selling model with 1,812,800 units, followed by the Aqua with 1,154,500 units, the Prius α with 446,400, and the Prius plug-in with 22,100. Cumulative sales of Honda's hybrid vehicles since November 1999 reached 25,239 units by January 2009, and in March 2010, Honda announced that the new 2010 Insight broke through 100,000 sales in Japan in just one year after its introduction.

The Toyota Prius α was launched in Japan in May 2011.

Hybrid sales in Japan almost tripled in 2009 as compared to 2008 as a result of government incentives that included a scrappage program, tax breaks on hybrid vehicles and other low-emission cars and trucks, and a higher levy on gasoline that rose prices in the order of US$4.50. New hybrid car sales jumped from 94,259 in 2008 to 334,000 in 2009, and hybrid sales in 2009 represented around 10% of new vehicles sales in Japan. In contrast, the U.S. market share was 2.8% for the same year. These record sales allowed Japan to surpass the U.S. in total new hybrid sales, with the Japanese market representing almost half (48%) of the worldwide hybrid sales in 2009 while the U.S. market represented 42% of global sales. The Toyota Prius became the first hybrid to top annual new car sales in Japan with 208,876 units sold in 2009. The Insight ranked fifth in overall sales in 2009 with 93,283 units sold.

A total of 315,669 Priuses were sold domestically in 2010, making the Prius the country's best-selling vehicle for the second straight year. Also the Prius broke Japan's annual sales record for a single model for the first time in 20 years, surpassing the Toyota Corolla, which in 1990 set the previous sales record with 300,008 units. The Prius sold 252,528 units in 2011, becoming the best-selling vehicle for the third-consecutive year. This figure includes sales of the Prius α, launched in May 2011, and the Toyota Aqua, launched in December. Despite keeping to the top-selling spot, total Prius sales for 2011 were 20% lower than 2010 due partly to the disruptions caused by the March 2011 Tōhoku earthquake and tsunami, and also because government incentives for hybrid cars were scaled back. Nevertheless, during the 2011 Japanese fiscal year (April 1, 2011 through March 31, 2012), hybrid vehicles accounted for 16% of all new car sales in the country. In May 2012, hybrid sales reached a record market share of 19.7% of new car sales in the country, including kei cars. Sales were led by the conventional Prius followed by the Toyota Aqua. Also during this month, hybrid sales represented 25% of Honda sales and 46% of Toyota sales in the country.

The Toyota Aqua has been the top-selling new car in Japan for three years running, from 2013 to 2015.

The Toyota Aqua, released in December 2011, ranked as the second top-selling new car in Japan in 2012 after the conventional Prius. Totaling 262,367 units sold in 2013, the Aqua topped new car sales in Japan in 2013, including kei car sales. And with 233,209 units sold during 2014, down 11.1% from 2013, the Aqua was the top-selling new car in Japan for the second consecutive year. Again in 2015, with 215,525 units sold, down 7.6% from 2014, the Aqua ranked as the top-selling new car in Japan. The Toyota Aqua not only was the best-selling new car in Japan for three years running, from 2013 to 2015, but it is considered the most successful nameplate launch in the Japanese market of the last 20 years. In the first quarter of 2016, the Prius liftback surpassed the Aqua as the best-selling new car, the Prius ended 2016 as the best-selling standard car in the Japanese market with 248,258 units, followed by the Aqua with 168,208 units.

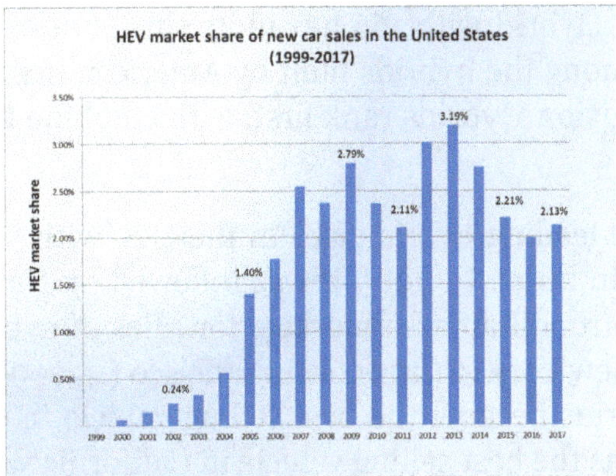

HEV market share of new car sales in the U.S. between 1999 and 2017.

The market of hybrid electric vehicles in the United States is the second largest in the world after Japan, and represents about 36% of the more than 11 million hybrid sold worldwide through April 2016. Cumulative sales passed 2 million units in May 2011, the 3 million mark in October 2013, and 4 million in April 2016. Sales of hybrid vehicles in the U.S. began to decline following the financial crisis of 2007–08, and after a short recovery, began to decline again in 2014 due to low gasoline prices. Hybrid sales in the American market achieved its highest market share ever in 2013, capturing 3.19% of new car sales that year. At the end of 2015 the hybrid take rate had fallen to 2.21%, dropped to 1.99% in 2016, and slightly recovered to 2.13% in 2017.

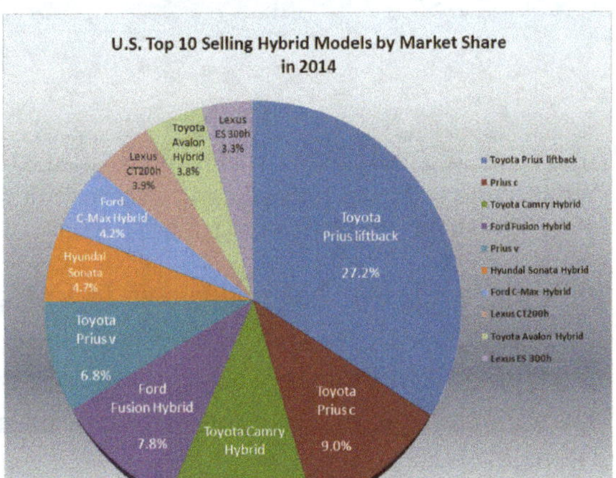

U.S. best selling HEV models in 2014 by market share.

Since their inception in 1999, a total of 4,058,258 hybrid electric automobiles and sport utility vehicles have been sold in the country through May 2016. The Toyota Prius family is the market leader with 1,932,805 units sold through April 2016, representing a 48.0% market share of total hybrid sales. The conventional Toyota Prius is the top-selling hybrid model, with 1,643,000 units sold through April 2016, accounting for 40.8% of all hybrids sold in the U.S. since inception. The United States accounted for 44.7% of Toyota Motor Company global hybrid sales through April 2016.

As of April 2016, sales of the conventional Prius are followed by the Toyota Camry Hybrid, with 345,640 units sold since 2006, the Honda Civic Hybrid, with cumulative sales of 234,610 vehicles since 2002, the Ford Fusion Hybrid with 166,341 units since 2009, and the Toyota Prius c with 165,075 units since 2012. Among the hybrids built by American manufacturers, the siblings Lincoln MKZ/Mercury Milan/Fusion Hybrids rank first, with combined sales of 200,899 units since 2009 through 2015.

California has been the state leading hybrid sales in the U.S. with 55,553 vehicles sold in 2009, 74,932 in 2008, and 91,417 in 2007. In 2009 it was followed by New York (15,438) and Florida (14,949). The top 5 U. S. metropolitan area markets for sales of hybrid electric vehicles in 2009 were Los Angeles (26,677), New York (21,193), San Francisco (15,799), Washington, D.C. (11,595), and Chicago (8,990). The Prius became the best selling vehicle in California in 2012. Again in 2013, the Prius nameplate was the best selling vehicle in California with 69,728 units sold.

European Market

The Toyota Auris Hybrid is the top-selling hybrid electric vehicle in Europe with 427,600 units through 2017.

As of April 2016, about 1.5 million hybrids cars have been sold in Europe since their introduction, representing around 13.6% of cumulative global hybrid sales. Of these, Toyota and Lexus hybrids totaled around 1.1 million units (74%). PSA Peugeot Citroën had sold over 50,000 diesel-powered hybrids in Europe through December 2013. In 2014, one-fourth of all new vehicles sold by Toyota in the European Union were hybrid-electric. The top-selling hybrid markets in 2015 were France, followed by the UK, Italy, Germany, Spain, the Netherlands, and Norway.

As of December 2014, hybrid vehicles accounted for 1.4% of new passenger car registrations in the EU Member States, up from 1.1% in 2012. The Netherlands is the leading country within the Union states with a market share of 3.7% of total passenger car sales in 2014, though fewer hybrid vehicles were sold in the Dutch market in 2014 than in previous years. Accounting for all countries in the continent, Norway is the segment leader, with a market share of 6.9% in 2014.

Sales of hybrids in Europe went up from around 9,000 units in 2004 to 39,880 in 2006, with Toyota accounting for 91% of hybrid sales and Honda with 3,410 units sold that year. Cumulative sales of Toyota hybrids since 2000 reached 69,674 units in 2006, while Honda hybrid sales reached over 8,000 units. By January 2009, Honda had sold 35,149 hybrids in Europe, of which 34,757 were Honda Civic Hybrids. During 2008 combined sales of Toyota and Lexus hybrids in Europe were 57,819 units, representing 5.2% of total Toyota sales in the region.

Toyota sales were led by Prius with 41,495 units. Cumulative sales of the Toyota Prius reached 100,000 units in 2008 and the 200,000 mark was reached in July 2010. The UK has been one of the leading European markets for the Prius since its inception, with 20% of Prius sales in Europe by 2010.

The Toyota Yaris Hybrid, released in 2012, is available in Europe in lieu of the Prius c.

Toyota's European hybrid sales reached 70,529 vehicles in 2010, including sales of 15,237 Toyota Auris Hybrids. Sales reached 84,839 units in 2011, including 59,161 Toyota and 25,678 Lexus hybrid vehicles. The Auris hybrid sold 32,725 units in 2011. Lexus hybrids made up 85% of total sales in Western Europe in 2011. Toyota and Lexus hybrids represented 10% percent of Toyota's European new car sales in 2011. TMC share of hybrid sales out of the company's total European sales climbed from 13% in 2012 to 20% during the first 11 months of 2013.

During the first nine months of 2013, over 118,000 hybrids were sold in Western Europe representing a 1.4% market share of new car sales in the region. A total of 192,664 hybrid cars were sold in the European Union and EFTA countries in 2014. Sales increased 21.5% in 2015, with 234,170 units sold. In 2015, gasoline-powered hybrids represented 91.6% of total hybrid registrations. The top-selling model in 2015 was the Toyota Auris Hybrid, with 75,810 units, up 13.0% from 2014; followed by the Yaris Hybrid with 68,902 units, up 22% from 2014. Seven of the top ten hybrids models sold in 2015 were from either Toyota or the Lexus brand. Toyota achieved record hybrid sales in 2015 with 201,500 units delivered. Hybrid registrations in the European Union and EFTA countries totaled 74,796 units during the first quarter of 2016, up 29.7% from the same quarter the previous year.

Cumulative TMC sales since the Prius introduction in Europe in 2000 passed the one million unit milestone in November 2015. As of December 2017, the top-selling Toyota hybrids were the Auris Hybrid (427,600), the Yaris Hybrid (388,900), and the conventional Prius (299,100). The top-selling Lexus models are the Lexus RX 400h/RX 450h with 111,100 units, and the Lexus CT 200h with 78,100 units.

Since 2006 hybrid car registrations in the UK totaled 257,404 units up to April 2016, including 11,679 diesel-electric hybrids, which were introduced in 2011. The market share of the British hybrid segment climbed from 1.1% in 2010, to 1.2% in 2012, and achieved 1.5% of new car registrations in 2014.

Since 2000, when the Prius was launched in the UK, 100,000 Toyota hybrids had been sold by May 2014, and almost 50,000 Lexus models since the introduction of the RX 400h in 2005. Honda had sold in the UK more than 22,000 hybrid cars through December 2011 since the Insight was launched in the country in 2000. After 15 years since the launch of the Prius in the British market, combined sales of Toyota and Lexus hybrids reached the 200,000 unit milestone in November 2015.

A total of 37,215 hybrids were registered in 2014, and while petrol-electric hybrids increased 32.6% from 2013, diesel-electric hybrids declined 12.6%. Hybrid registrations totaled a record of 44,580 units in 2015, consisting of 40,707 petrol-powered hybrids and 3,873 powered by diesel; the latter experienced a 36.3% increase from 2014, while petrol-powered hybrid grew by 18.1%. The hybrid segment market shared reached 1.69% of new car registrations in the UK that year.

A total of 165,915 hybrid cars have been registered in France between 2007 and 2014, including 33,547 diesel-powered hybrids. French registrations account plug-in hybrid together with conventional hybrids. Among EU Member States, France had the second largest hybrid market share in 2014, with 2.3% of new car sales, down from 2.6% in 2013.

Diesel hybrid technology, introduced by PSA Peugeot Citroën with the HYbrid4 system in 2011, represented 20.2% of the hybrid car stock sold in France between 2011 and 2014. Among the 13,340 units registered in 2011, the top-selling models in the French market were the Toyota Auris (4,740 units), the Prius (2,429 units), and the Honda Jazz Hybrid (1,857 units). The diesel-powered Peugeot 3008 HYbrid4, launched in late 2011, sold 401 units. Toyota led hybrid sales in the French market in 2013 with 27,536 registrations of its Yaris, Auris and Prius models, followed by the PSA group with 13,400 registrations. During 2014, a total of 42,813 hybrid cars and vans were registered, down 8.5% from 2013. Of these, 9,518 were diesel-electric hybrids, down 31.9% from 13,986 units a year earlier, while registrations of gasoline-electric hybrids were up 1.5%. The top-selling models in 2014 were the Toyota Yaris Hybrid with 12,819 units, Toyota Auris with 10,595 and the Peugeot 3008 with 4,189 units. Hybrid registrations in 2014 included 1,519 plug-in hybrids, with sales led by the Mitsubishi Outlander P-HEV, with 820 units.

As of 31 December 2015, hybrid car registrations totaled 131,011 units, up 11.7% from 117,259 a year earlier. By the end of 2009 there were about 39,300 hybrid cars registered in the Netherlands, up from 23,000 the previous year. Most of the registered hybrid cars belonged to corporate fleets due to tax incentives established in the country in 2008. During the first eight months of 2013, around 65% of TMC cars sold in the Netherlands have been hybrids, with the technology particularly popular among fleet owners and taxi drivers. Following the same market trend as in 2014, more plug-in hybrids were registered in 2015 (41,226) in the country than conventional hybrids (13,752).

As a result of the tax incentives, the country has had for several years the highest hybrid market share among EU Member States. Hybrid sales climbed from 0.7% in 2006 and 2007 to 2.4% in 2008, and reached 4.2% in 2009. Due to the financial crisis of 2007–08, the market fell for two years to 2.7% in 2011, but recovered to 4.5% in 2012. As fewer hybrid vehicles were sold in the Dutch market in 2014 than in previous years, the hybrid segment market share fell to 3.7% of total passenger car sales in 2014. The sales decline is due to a change in the national vehicle taxation scheme. As of 2014, Japan (~20%) and Norway (6.9%) are the only countries with a higher market share than the Netherlands.

As of January 2016, there where were 130,365 hybrid cars registered in Germany, up from 85,575 on the roads in January 1, 2014, and 47,642 vehicles on January 1, 2012. Hybrid car registrations totaled a record of 24,963 units in 2013, and declined to 22,908 in 2014, and to 22,529 in 2015. The German hybrid market share climbed from 0.3% in 2010, to 0.8% in 2013, and declined 0.7% of new car registrations in 2014.

A total of 10,350 hybrid cars were registered in Spain in 2011, up 22% from 2010 sales. The top-selling hybrids were the Toyota Prius, Toyota Auris HSD and the Lexus CT 200h, which together represented 83,2% of new hybrid car sales in the country. During 2012 hybrid sales remained almost constant with 10,030 units sold, representing 1.44% of new passenger cars sales that year. The top-selling car was the Prius with 3,969 units, followed by the Auris HSD (2,234) and the Lexus CT 200h (1,244). Combined sales of Toyota and Lexus models represented 89.15% of hybrid sales in the Spanish market in 2012. Hybrid sales in 2013 increased 1.72% from 2012, with 10,294 units registered. The Toyota Auris HSD was the top-selling hybrid with 3,644 units, followed by the Prius (2,378) and the Yaris Hybrid (1,587).

Technology

The varieties of hybrid electric designs can be differentiated by the structure of the hybrid vehicle drivetrain, the fuel type, and the mode of operation.

In 2007, several automobile manufacturers announced that future vehicles will use aspects of hybrid electric technology to reduce fuel consumption without the use of the hybrid drivetrain. Regenerative braking can be used to recapture energy and stored to power electrical accessories, such as air conditioning. Shutting down the engine at idle can also be used to reduce fuel consumption and reduce emissions without the addition of a hybrid drivetrain. In both cases, some of the advantages of hybrid electric technology are gained while additional cost and weight may be limited to the addition of larger batteries and starter motors. There is no standard terminology for such vehicles, although they may be termed mild hybrids.

Engines and Fuel Sources

Fossil Fuels

Free-piston engines could be used to generate electricity as efficiently as, and less expensively than, fuel cells.

Gasoline

Gasoline engines are used in most hybrid electric designs and will likely remain dominant for the foreseeable future. While petroleum-derived gasoline is the primary fuel, it is possible to mix in varying levels of ethanol created from renewable energy sources. Like most modern ICE powered vehicles, HEVs can typically use up to about 15% bioethanol. Manufacturers may move to flexible fuel engines, which would increase allowable ratios, but no plans are in place at present.

Diesel

Diesel-electric HEVs use a diesel engine for power generation. Diesels have advantages when delivering constant power for long periods of time, suffering less wear while operating at higher efficiency. The diesel engine's high torque, combined with hybrid technology, may offer substantially improved mileage. Most diesel vehicles can use 100% pure biofuels (biodiesel), so they can use but do not need petroleum at all for fuel (although mixes of biofuel and petroleum are more common). If diesel-electric HEVs were in use, this benefit would likely also apply. Diesel-electric hybrid drivetrains have begun to appear in commercial vehicles (particularly buses); as of 2007, no light duty diesel-electric hybrid passenger cars are available, although prototypes exist. Peugeot is expected to produce a diesel-electric hybrid version of its 308 in late 2008 for the European market.

PSA Peugeot Citroën has unveiled two demonstrator vehicles featuring a diesel-electric hybrid drivetrain: the Peugeot 307, Citroën C4 Hybride HDi and Citroën C-Cactus. Volkswagen made a prototype diesel-electric hybrid car that achieved 2 L/100 km (140 mpgimp; 120 mpgUS) fuel economy, but has yet to sell a hybrid vehicle. General Motors has been testing the Opel Astra Diesel Hybrid. There have been no concrete dates suggested for these vehicles, but press statements have suggested production vehicles would not appear before 2009.

At the Frankfurt Motor Show in September 2009 both Mercedes and BMW displayed diesel-electric hybrids.

Robert Bosch GmbH is supplying hybrid diesel-electric technology to diverse automakers and models, including the Peugeot 308.

So far, production diesel-electric engines have mostly[vague] appeared in mass transit buses.

FedEx, along with Eaton Corp. in the USA and Iveco in Europe, has begun deploying a small fleet of Hybrid diesel electric delivery trucks. As of October 2007, Fedex operates more than 100 diesel electric hybrids in North America, Asia and Europe.

Liquefied Petroleum Gas

Hyundai Elantra LPI Hybrid.

Hyundai introduced in 2009 the Hyundai Elantra LPI Hybrid, which is the first mass production hybrid electric vehicle to run on liquefied petroleum gas (LPG).

Hydrogen

Hydrogen can be used in cars in two ways: a source of combustible heat, or a source of electrons for an electric motor. The burning of hydrogen is not being developed in practical terms; it is the hydrogen fuel-cell electric vehicle (HFEV) which is garnering all the attention. Hydrogen fuel cells create electricity fed into an electric motor to drives the wheels. Hydrogen is not burned, but it is consumed. This means molecular hydrogen, H2, is combined with oxygen to form water. $2H_2$ $(4e^-)$ + O_2 --> $2H_2O$ $(4e^-)$. The molecular hydrogen and oxygen's mutual affinity drives the fuel cell to separate the electrons from the hydrogen, to use them to power the electric motor, and to return them to the ionized water molecules that were formed when the electron-depleted hydrogen combined with the oxygen in the fuel cell. Recalling that a hydrogen atom is nothing more than a proton and an electron; in essence, the motor is driven by the proton's atomic attraction to the oxygen nucleus, and the electron's attraction to the ionized water molecule.

An HFEV is an all-electric car featuring an open-source battery in the form of a hydrogen tank and the atmosphere. HFEVs may also comprise closed-cell batteries for the purpose of power storage from regenerative braking, but this does not change the source of the motivation. It implies the HFEV is an electric car with two types of batteries. Since HFEVs are purely electric, and do not contain any type of heat engine, they are not hybrids.

Biofuels

The Ford Escape Hybrid was the first hybrid electric vehicle with a flex-fuel engine capable of running on E85 fuel.

Demonstration Ford Escape E85 flex-fuel plug-in hybrid.

Hybrid vehicles might use an internal combustion engine running on biofuels, such as a flexible-fuel engine running on ethanol or engines running on biodiesel. In 2007 Ford produced 20 demonstration Escape Hybrid E85s for real-world testing in fleets in the U.S. Also as a demonstration project, Ford delivered in 2008 the first flexible-fuel plug-in hybrid SUV to the U.S. Department of Energy (DOE), a Ford Escape Plug-in Hybrid, capable of running on gasoline or E85.

The Chevrolet Volt plug-in hybrid electric vehicle would be the first commercially available flex-fuel plug-in hybrid capable of adapting the propulsion to the biofuels used in several world markets such as the ethanol blend E85 in the U.S., or E100 in Brazil, or biodiesel in Sweden. The Volt will be E85 flex-fuel capable about a year after its introduction.

Electric Machines

In split path vehicles (Toyota, Ford, GM, Chrysler) there are two electrical machines, one of which functions as a motor primarily, and the other functions as a generator primarily. One of the primary requirements of these machines is that they are very efficient, as the electrical portion of the energy must be converted from the engine to the generator, through two inverters, through the motor again and then to the wheels.

Most of the electric machines used in hybrid vehicles are brushless DC motors (BLDC). Specifically, they are of a type called an interior permanent magnet (IPM) machine (or motor). These machines are wound similarly to the induction motors found in a typical home, but (for high efficiency) use very strong rare-earth magnets in the rotor. These magnets contain neodymium, iron and boron, and are therefore called Neodymium magnets.

The price of Neodymium went through a price bubble due to Chinese export restriction in 2010–11, rising from $50/kg at the start of 2010 to $500/kg by the summer of 2011. This resulted in 'demand destruction' as many producers quickly turned to substituting induction motors in their cars to defend their production line. This was in spite of such motors inferior 'power to weight' ratio attributes significantly impacting all but the most powerful (energy guzzling) motor sizes, e.g. those used in the Tesla. As of April 2014 there are other non-Chinese producers of Neodymium and its price/kg is not much more than it was in 2010. Cutting edge U.K. motors that are now being produced are using Neodymium Permanent Magnet technology. As security of supply returns, it is certain that there will accordingly be a return to superior motor designs that NdFeB Permanent Magnets enable.

Design Considerations

In some cases, manufacturers are producing HEVs that use the added energy provided by the hybrid systems to give vehicles a power boost, rather than significantly improved fuel efficiency compared to their traditional counterparts. The trade-off between added performance and improved fuel efficiency is partly controlled by the software within the hybrid system and partly the result of the engine, battery and motor size. In the future, manufacturers may provide HEV owners with the ability to partially control this balance (fuel efficiency vs. added performance) as they wish, through a user-controlled setting. Toyota announced in January, 2006 that it was considering a "high-efficiency" button.

Conversion Kits

One can buy a stock hybrid or convert a stock petroleum car to a hybrid electric vehicle using an aftermarket hybrid kit.

The advantages of HEVs include:

- Oil consumption is less than that of conventional vehicles.
- Carbon-based emission is lower, which makes HEVs more eco-friendly. This also helps conserve non-renewable resources like petroleum products.
- Maintenance costs are lower than those of conventional vehicles.
- With the electric motor taking charge of the engine during long travels, more mileage can be achieved with HEVs compared to other types of vehicles.

Plug-in hybrid electric vehicle

Plug-in hybrid electric vehicles–known as PHEVs–combine a gasoline or diesel engine with an electric motor and a large rechargeable battery. Unlike conventional hybrids, PHEVS can be plugged-in and recharged from an outlet, allowing them to drive extended distances using just electricity. When the battery is emptied, the conventional engine turns on and the vehicle operates as a conventional, non-plug-in hybrid.

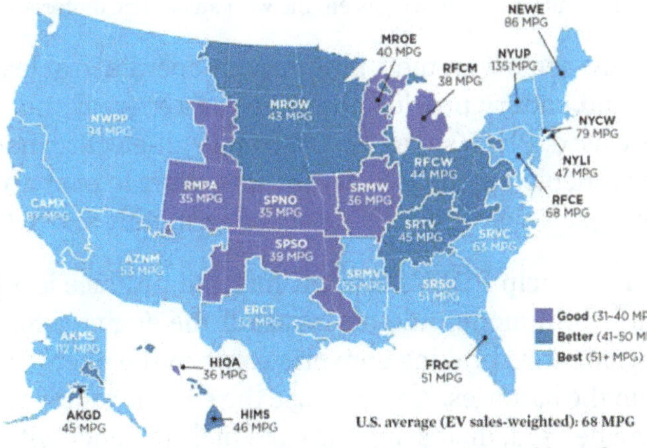

Driving on electricity offers huge environmental benefits.

Because they can run on electricity from the grid—and because electricity is often a cleaner energy source than gasoline or diesel—plug-in hybrids can produce significantly less global warming pollution than their gas-only counterparts. They don't emit any tailpipe pollution when driving on electricity, and they gain fuel efficiency benefits from having an electric motor and battery. Since they use less gas, they also cost less to fuel: driving a PHEV can save hundreds of dollars a year in gasoline and diesel costs.

To take advantage of plug-in hybrid vehicles, drivers need access to parking and a place to plug-in, though a regular 120V outlet is usually sufficient. And since most PHEVs are passenger cars, prospective buyers shouldn't regularly require space for more than five occupants, and they shouldn't need to tow.

Plug-in Hybrid Features

PHEVs combine the fuel-savings benefits of hybrids with the all-electric capabilities of battery-electric or fuel-cell vehicles.

Plug-in hybrid vehicles combine a gas engine with an electric motor and battery.

While not all models work the same way, most plug-ins can operate in at least two modes: "all-electric," in which the motor and battery provide all of the car's energy; and "hybrid," in which both electricity and gasoline are used. PHEVs typically startup in all-electric mode, running on electricity until their battery pack is depleted: ranges vary from 10 miles to over 40. Certain models switch to hybrid mode when they reach highway cruising speed, generally above 60 or 70 miles per hour.

The electric motor and battery help PHEVs use less fuel and produce less pollution than conventional cars, even when in hybrid mode. Idle-off turns off the engine while idling at stoplights or in traffic, saving fuel. Regenerative braking converts some of the energy lost during braking into usable electricity, stored in the batteries. And because the electric motor supplements the engine's power, smaller engines can be used, increasing the car's fuel efficiency without compromising performance.

Different models of plug-in hybrid electric vehicles may also have different drivetrains, the mechanical components that deliver power to the driving wheels.

Different Kinds of PHEVs

There are two basic PHEV configurations:

- Series PHEVs, also called Extended Range Electric Vehicles. Only the electric motor turns the wheels; the gasoline engine only generates electricity. Series PHEVs can run solely on electricity until the battery needs recharging. The gasoline engine will then generate the electricity needed to power the electric motor. For shorter trips, these vehicles might use no gasoline at all.
- Parallel or Blended PHEVs. Both the engine and electric motor are mechanically connected to the wheels, and both propel the vehicle under most driving conditions. Electric-only operation usually occurs only at low speeds.

PHEVs also have varied battery capacities, allowing some to travel farther on electricity than others. PHEV fuel economy, like that of electric vehicles and regular hybrids, can be sensitive to driving style, driving conditions, and accessory use.

Working of PHEVs

In addition to plugging into an outside electric power source, PHEV batteries can be charged by an internal combustion engine or regenerative braking. (During braking, the electric motor acts as a generator, using the energy to charge the battery.) The electric motor supplements the engine's power; as a result, smaller engines can be used, increasing the car's fuel efficiency without compromising performance.

PHEVs typically start up in all-electric mode and operate on electricity until their battery pack is depleted. Some models shift to hybrid mode when they reach highway cruising speed, generally above 60 or 70 miles per hour. Once the battery is empty, the engine takes over and the vehicle operates as a conventional, non-plug-in hybrid.

There are two basic plug-in hybrid configurations, according to the DoE:

In series plug-in hybrids, or extended-range electric vehicles, only the electric motor turns the wheels and the gasoline engine generates electricity. Series plug-ins can run on electricity alone until the battery runs down. The gasoline engine then generates electricity to power the electric motor. These vehicles might use no gasoline at all for short trips.

In parallel, or blended plug-in hybrids, both the engine and electric motor are connected to the wheels and propel the vehicle under most driving conditions. The electric-only operation usually occurs only at low speeds.

PHEV Costs and Benefits

Although the electric motor and battery help plug-in hybrids use less fuel and produce less pollution than conventional cars, even when in hybrid mode, fuel consumption depends on the distance driven between battery charges. If the vehicle is never plugged in to charge, fuel economy will be about the same as a similarly sized hybrid electric vehicle. If the vehicle is driven a shorter distance than its all-electric range and plugged in to charge between trips, it may use only electric power.

Plug-in hybrids have a significantly lower carbon footprint than their gas-only counterparts for two reasons: PHEVs can run on electricity from the power grid, and electricity is typically cleaner than gasoline or diesel fuel. Plug-in hybrids don't emit tailpipe pollution when driving on electricity; plus, having an electric motor and battery is fuel-efficient.

Although plug-in hybrids typically emit less greenhouse gas than conventional vehicles, the amount generated depends partly on how the electricity is produced. For example, nuclear and hydroelectric plants are cleaner than coal-fired power plants, according to the DoE.

Additionally, although using electricity is usually cheaper than using gasoline, fuel savings may not always offset the higher PHEV cost (approximately $4,000 to $8,000), according to the DoE. Fuel savings depend on the vehicle, how many miles it operates using electricity and fuel costs. Plug-in hybrid cars purchased in 2010 or later may be eligible for federal tax credits of up to $7,500; state and local incentives may also apply.

Re-charging a PHEV is time-consuming – several hours using a 120-volt outlet and 1 to 4 hours using a 240-volt charger. A "fast charge" to 80 percent capacity may take as little as 30 minutes, according to the DoE.

Challenges

Higher Vehicle Costs, Lower Fuel Costs. PHEVs will likely cost $1,000 to $7,000 (USD) more than comparable non-plug-in hybrids. Fueling a PHEV will cost less because the cost of electricity is much lower than the cost of gasoline per mile, but fuel savings will not entirely offset the increased vehicle cost. Many governments provide incentives to consumers for qualifying PHEVs.

Charging Takes Time. Charging the battery typically takes several hours, but a "quick charge" to 80% capacity may take as little as 30 minutes. However, PHEVs can be driven without being plugged in. They can be fueled solely with gasoline but will not achieve maximum range or fuel economy without charging.

Measuring Fuel Economy. This will be a challenge because PHEVs run on both electricity and liquid fuel, and it is unclear what percentage of time each will be used during the average driver's daily travel.

Hybrid Electric Truck

Hybrid-electric trucks achieve increased fuel efficiency by combining a conventional internal combustion engine with an electric motor, batteries, and braking-energy capture (known as regenerative braking).

- Fuel-consumption reductions of 20 to 35 percent are possible with current hybrid truck technology, according to a review by the National Academy of Sciences.
- Commercial production of the first hybrid delivery truck began in 2007. Today several manufacturers offer hybrid models, and thousands of hybrid systems are being used in applications ranging from public-transit buses to package- and beverage-delivery trucks.

Hybrid trucks use 20 to 50 percent less diesel than conventional vehicles do, depending on how they're used, and that adds up: FedEx calculated that its fleet of 170 hybrid–electric trucks has racked up 3.5 million miles (5.6 million kilometers) of service. The 200 hybrid trucks United Parcel Service (UPS) will add to its fleet later this year are expected to save 176,000 gallons (665,000 liters) of fuel per year, which is the equivalent of taking 100 conventional package delivery trucks off the road. Coca-Cola Enterprises is the most avid consumer of hybrid trucks in the U.S., deploying 327 hybrids out of a total fleet of 30,000. Depending on how its driven, a typical medium-duty hybrid delivery truck from Peterbilt will average 12 miles per gallon (5.1 kilometers per liter), versus nine mpg (3.8 kpl) for the conventional version of the same truck, according to Peterbilt. The largest trucks on the road average six mpg

(2.5 kpl), so the increase to eight mpg (3.4 kpl) achieved by Peterbilt's hybrid technology is significant.

Hybrid Electric Bus

HEBs use both an internal combustion engine (ICE), which is usually diesel powered, and an electric motor (EM) to power the vehicle. They are the most common type of electric bus in operation globally, and continue to be the most purchased type of new electric bus.

Technology Overview

- A hybrid-electric bus is powered by an electric motor, and a smaller than normal conventional internal combustion engine. To achieve heightened efficiency, the battery powered electric motor allows the combustion engine to operate at periods of maximum efficiency. The major components of a hybrid drive system include an internal combustion engine (ICE), a generator, a battery pack, and an electric motor. The most commonly used batteries in hybrid vehicles are lead-acid or nickel-metal hydride.
- Ultra low sulfur diesel is the most common fuel used to power the combustion engine in hybrids, although other fuels such as gasoline, compressed natural gas (CNG), liquid natural gas (LNG), biodiesel, and hydrogen have also been used.
- A majority of the hybrid buses in service are 40-foot buses. However, 22-foot shuttle buses and 60-foot articulated buses have also been deployed.
- The three major manufacturers of hybrid systems are GM-Allison Transmission, BAE Systems, and ISE Corporation. Ebus produces 22-foot shuttle buses.
- There are two different types of hybrid systems - parallel and series hybrids.
    - In a parallel hybrid bus, the combustion engine and the electric motor are connected to the transmission independently. The electric motor is designed to provide power during stop-and-go traffic while at highway speeds the vehicle is powered solely by the internal combustion engine. During acceleration, both the electric motor and the combustion engine power the transmission. In addition, through a process called regenerative breaking, energy lost due to braking is recovered and utilized to charge the battery.
    - A series hybrid bus is exclusively propelled by the electric motor. In a series hybrid bus, the internal combustion engine (ICE) is connected to an electric generator which converts the energy produced by the ICE into electric power. This electricity powers a motor which turns the wheels of the vehicle. The generator also recharges a battery pack which provides supplemental power to the motor. Since the ICE is not connected to the wheels, it can operate at an optimum rate and can even be switched off for short periods of time for a temporary all-electric operation of the bus.
- Parallel hybrids have greater fuel efficiency than series hybrids at constant high speeds (highway driving) while the series hybrids fare better in stop-and-go urban driving.

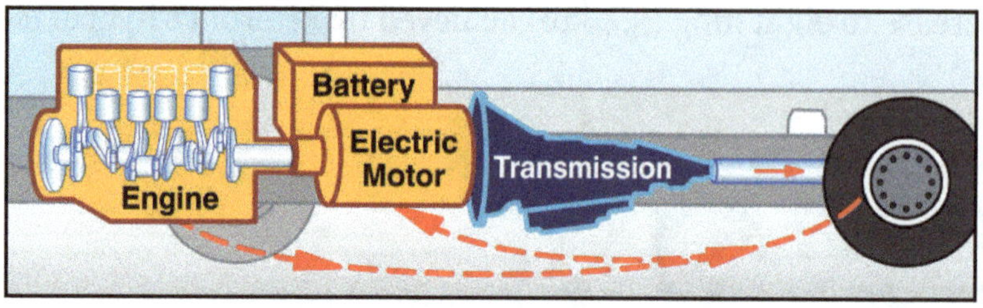

Parallel hybrids electric configuration

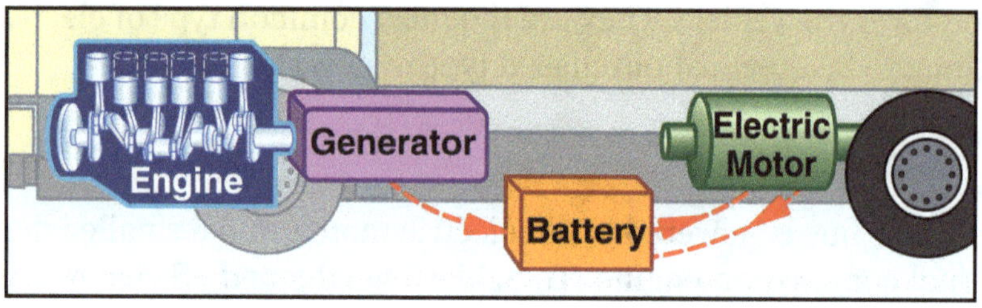

Series hybrids electric configuration

Benefits of Hybrid Buses

Hybrid buses offer a wide range of benefits including significantly lower emissions, increased efficiency, and decreased maintenance costs.

Reduced Emissions

Hybrid buses are estimated to cut emissions by as much as 75 percent when compared to conventional diesel buses. The emissions reductions are a function of the electric drive, ultra low-sulfur diesel (ULSD) fuel use in conjunction with particulate trap technology and improved fuel economy from the hybrid system.

- In-use testing and industry reports indicate that particulate matter (PM) emissions from hybrid buses equipped with particulate matter filters are almost 90 percent lower than a conventional diesel bus without a particulate filter.
- According to a study conducted by the Northeast Advanced Vehicle Coalition (NAVC) to demonstrate energy efficiency and emissions performance of diesel-electric hybrid buses, nitrogen oxides (NOx) emissions for diesel hybrids were 30 to 40 percent lower than conventional diesel vehicles.
- Diesel hybrid buses exhibited the lowest carbon monoxide (CO) emissions of any of the buses tested including CNG and conventional diesel.
- Hybrids demonstrated lower greenhouse gas emissions than those of a conventional diesel bus or CNG powered bus as a result of improved fuel economy.

A separate evaluation by New York City Transit (NYCT) reports significant emissions reductions for the Orion VII buses equipped with BAE Systems' series hybrid drive. In the chart below, emissions from the hybrid bus are compared with the emissions from a conventional diesel bus equipped with a diesel particulate filter and a comparable CNG bus.

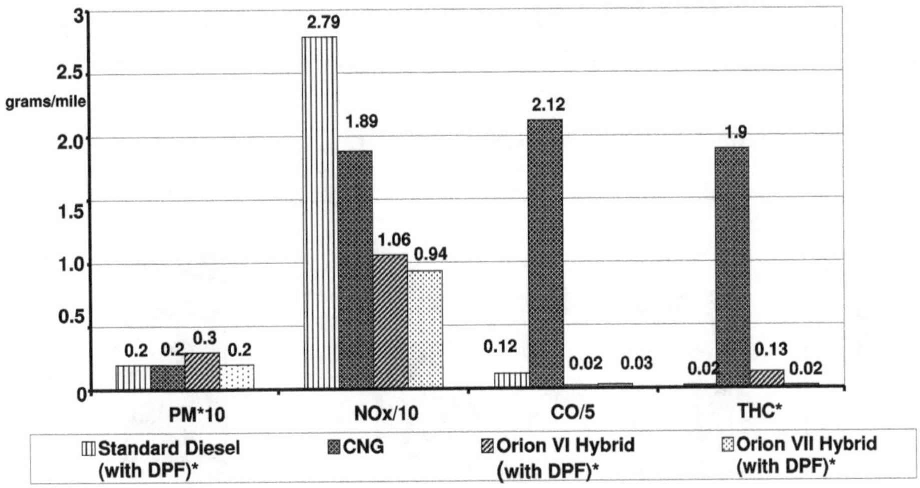

- Hybrid buses typically exhibit performance levels comparable to their non-hybrid counterparts. Transit agencies report that acceleration in hybrid-electric buses is smoother and faster due to the increased low-end torque characteristics of electric motors.
- Quieter Operation: Although there has been little systematic testing of hybrid buses for noise emissions, anecdotal evidence from bus drivers and passengers suggests that hybrid buses offer a quieter ride when compared to conventional diesel buses.
- Reduced Costs: The hybrid buses are expected to have lower maintenance costs due to reduced stress and maintenance on mechanical components such as brake linings, which may extend brake life by 50 – 100 percent. In addition, the electric drive has fewer parts, therefore requiring less maintenance than a traditional transmission.
- Increased fuel efficiency: Results from the National Renewable Energy Laboratory's (NREL) evaluations show significant increases in fuel economy compared to conventional diesel buses. NREL evaluated 10 CNG 40-foot Orion VII buses (Model Year 2002) equipped with Detroit Diesel Corporation's Series 50G CNG engine and 10 hybrid 40-foot Orion VII buses (Model Year 2002) fitted with BAE Systems series HybriDrive propulsion system. These buses were compared with nine standard diesel Orion V buses (Model Years 1994 and 1999) for bus usage, fuel economy and miles between maintenance calls. The evaluated buses were part of a 260 CNG bus order ($319,000 each) and a 125 diesel-electric hybrid bus order ($385,000 each) in 2003 and 2004.

Results from the evaluation indicate that the hybrid buses offered on average, a 37 percent improvement in fuel economy over standard diesel buses. It is important to note that the comparison to the standard diesel bus is limited because the engine fitted in these older buses (DDC Series 50) is no longer available for transit application. Further, neither the baseline standard diesel buses nor the hybrids were fitted with exhaust gas recirculation (EGR) for emission control. EGR systems tend to lower fuel efficiency. The fuel economy improvements for hybrids fluctuated between 26 – 52 percent higher (compared to diesel buses). Typically fuel economy is expected to decrease when the hybrid bus is operated in the summer months due to increased energy demand by vehicle accessories such as air conditioning.

In comparison to the CNG buses, the hybrids improved fuel efficiency on average by 88 percent. The DDC Series 50 CNG engines used in the evaluation are no longer available for transit use. In comparison to diesel buses, CNG buses demonstrated 25 percent lower fuel economy.

Average Fuel Economy Comparisons

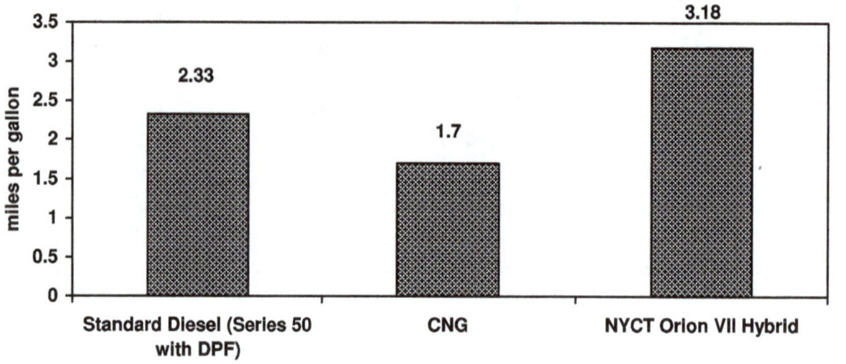

In conclusion, in-use testing of hybrid buses (both parallel and series, 40-foot and 60-foot configurations) reveals that they are more fuel efficient than conventional diesel buses, with fuel economy gains ranging from 28-48 percent depending on the type of hybrid and operating conditions.

Fuel is a huge expense for transit fleets, representing nearly \$700 million annually, or nearly three percent of operating costs. With rising diesel costs, increased fuel efficiency offered by the hybrids will result in substantial cost savings.

Hybrid Bus Cost Barriers

Currently hybrid buses carry a large price premium over conventional diesel buses. The average price of a 40-foot hybrid bus typically ranges from \$450,000 - \$550,000 when compared to \$280,000 - \$300,000 for a conventional diesel bus. The price variation in hybrids is due to the order volumes and individual specifications of transit agencies.

However the price differential for hybrids can be offset by various federal incentives and grant programs. Assuming that a standard diesel bus costs \$300,000 and a hybrid bus costs \$500,000, the incremental cost of purchasing a hybrid is \$200,000. The federal Clean Fuels Grant Program covers 90 percent of the incremental cost of alternative fuel buses, including hybrids. In addition, the Federal Transit Administration (FTA) covers up to 80 percent of the purchase price of a standard diesel bus.

Therefore a transit agency's share of matching funds for a hybrid is \$80,000 when compared to \$60,000 for a conventional diesel. This translates to an incremental cost of only \$20,000 to purchase a hybrid over a conventional diesel bus. As the technology matures and the market develops, the cost differential for hybrids is likely to decrease.

In addition, a year-long evaluation from NREL indicates that operational costs for hybrids are 15 percent lower than conventional diesel buses. According to their evaluation, King County Transit's 60-foot articulated New Flyer buses equipped with GM-Allison's parallel hybrid drive cost less to operate and maintain when compared to regular diesel buses. These cost savings, based on a diesel fuel price of $1.98, are likely to increase even further as diesel fuel prices continue to sky rocket.

King County Transit Parallel Hybrid Bus Operating Costs Comparison		
	Diesel (60-foot)	Diesel-electric Hybrid (60-foot)
Fuel ($1.98 per gallon)	$0.79/mi 2.50 mpg	$ 0.62/mi 3.17 mpg
Maintenance	$0.46/mi	$ 0.44/mi
Total	$1.25/mi	$1.06/mi

Current Applications

Hybrid bus deployment has grown tremendously in recent years. In 2006, there were more than 900 hybrid buses in regular servic e.i. more than 40 transit agencies in North America with several hundred on order.

Some of the largest hybrid fleets in the United States include New York City's fleet of 325 buses with an additional 500 vehicles on order, King County in Seattle which operates 214 parallel hybrid buses, and Washington DC which has placed an order for 100 diesel-electric hybrid buses. In addition, more than 15 states such as California, Connecticut, Indiana, Kentucky, Mississippi, New Mexico, Pennsylvania, Tennessee, and Texas to name a few, have deployed hybrid buses in their transit fleets with the several more looking to add hybrid buses to their fleets.

References

- Cox, C (2008). "Rare earth innovation: the silent shift to china". Herndon (VA): The Anchor House Inc. Archived from the original on 2008-04-21. Retrieved 2008-03-18
- Advantages-and-disadvantages-of-hybrid-cars: conserve-energy-future.com, Retrieved 28 May 2018
- Hybrid-electric-vehicle-hev-30892: techopedia.com, Retrieved 12 July 2018
- Lunn, J (3 October 2006). "Great western minerals" (PDF). London: Insigner Beaufort Equity Research. Archived from the original (PDF) on 2008-04-09. Retrieved 2008-03-18
- Electric-vehicles/series-vs-parallel-drivetrains, clean-vehicles: ucsusa.org, Retrieved 28 April 2018
- Plug-in-hybrid-electric-vehicle-PHEV: techtarget.com, Retrieved 28 June 2018
- MRC-Electric-Bus-Report-02082017: mrcagney.com, Retrieved 18 May 2018

3

Fuel Cell Electric Vehicles

A fuel cell electric vehicle is a kind of electric vehicle that uses a fuel cell in combination with a battery or a supercapacitor for powering its electric motor. The topics elaborated in this chapter on fuel cell, fuel cell vehicle, fuel cell forklift and fuel cell electric bus will help in providing a better perspective about the different types of fuel cell electric vehicles and their functioning.

A fuel cell is a device that generates electricity by a chemical reaction. Every fuel cell has two electrodes called, respectively, the anode and cathode. The reactions that produce electricity take place at the electrodes.

Every fuel cell also has an electrolyte, which carries electrically charged particles from one electrode to the other, and a catalyst, which speeds the reactions at the electrodes.

Hydrogen is the basic fuel, but fuel cells also require oxygen. One great appeal of fuel cells is that they generate electricity with very little pollution–much of the hydrogen and oxygen used in generating electricity ultimately combine to form a harmless byproduct, namely water.

One detail of terminology: a single fuel cell generates a tiny amount of direct current (DC) electricity. In practice, many fuel cells are usually assembled into a stack. Cell or stack, the principles are the same.

Working of Fuel Cells

The purpose of a fuel cell is to produce an electrical current that can be directed outside the cell to do work, such as powering an electric motor or illuminating a light bulb or a city. Because of the way electricity behaves, this current returns to the fuel cell, completing an electrical circuit. The chemical reactions that produce this current are the key to how a fuel cell works.

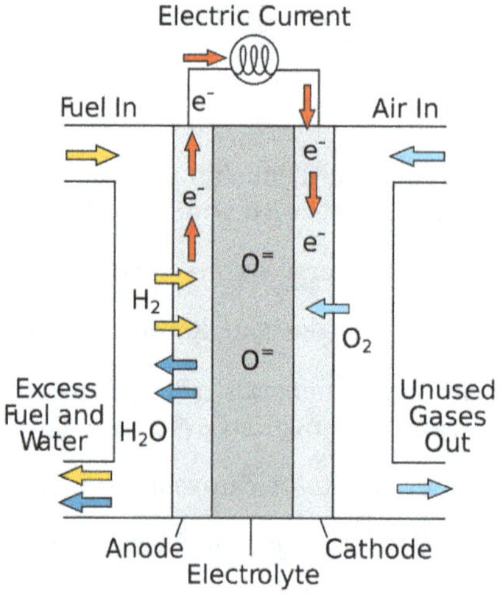

There are several kinds of fuel cells, and each operates a bit differently. But in general terms, hydrogen atoms enter a fuel cell at the anode where a chemical reaction strips them of their electrons. The hydrogen atoms are now "ionized," and carry a positive electrical charge. The negatively charged electrons provide the current through wires to do work. If alternating current (AC) is needed, the DC output of the fuel cell must be routed through a conversion device called an inverter.

Oxygen enters the fuel cell at the cathode and, in some cell types (like the one illustrated above), it there combines with electrons returning from the electrical circuit and hydrogen ions that have traveled through the electrolyte from the anode. In other cell types the oxygen picks up electrons and then travels through the electrolyte to the anode, where it combines with hydrogen ions.

The electrolyte plays a key role. It must permit only the appropriate ions to pass between the anode and cathode. If free electrons or other substances could travel through the electrolyte, they would disrupt the chemical reaction.

Whether they combine at anode or cathode, together hydrogen and oxygen form water, which drains from the cell. As long as a fuel cell is supplied with hydrogen and oxygen, it will generate electricity.

Even better, since fuel cells create electricity chemically, rather than by combustion, they are not subject to the thermodynamic laws that limit a conventional power plant. Therefore, fuel cells are more efficient in extracting energy from a fuel. Waste heat from some cells can also be harnessed, boosting system efficiency still further.

Types of Fuel Cells and Design

Fuel cells come in many varieties; however, they all work in the same general manner. They are made up of three adjacent segments: the anode, the electrolyte, and the cathode. Two chemical reactions occur at the interfaces of the three different segments. The net result of the two reactions is that fuel is consumed, water or carbon dioxide is created, and an electric current is created, which can be used to power electrical devices, normally referred to as the load.

At the anode a catalyst oxidizes the fuel, usually hydrogen, turning the fuel into a positively charged ion and a negatively charged electron. The electrolyte is a substance specifically designed so ions can pass through it, but the electrons cannot. The freed electrons travel through a wire creating the electric current. The ions travel through the electrolyte to the cathode. Once reaching the cathode, the ions are reunited with the electrons and the two react with a third chemical, usually oxygen, to create water or carbon dioxide.

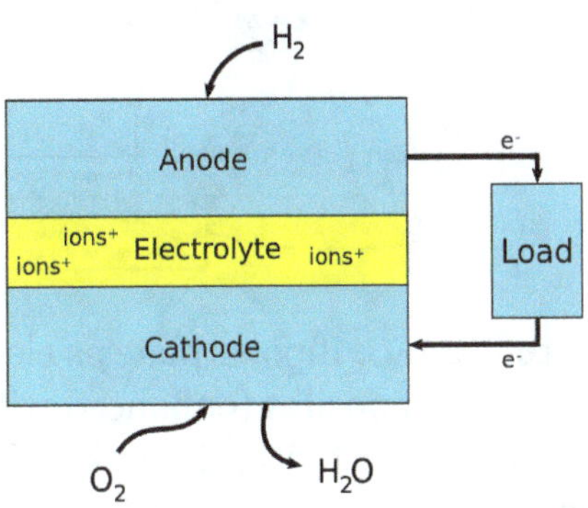

A block diagram of a fuel cell

Design features in a fuel cell include:

- The electrolyte substance, which usually defines the type of fuel cell, and can be made from a number of substances like potassium hydroxide, salt carbonates, and phosphoric acid.
- The fuel that is used. The most common fuel is hydrogen.
- The anode catalyst, usually fine platinum powder, breaks down the fuel into electrons and ions.
- The cathode catalyst, often nickel, converts ions into waste chemicals, with water being the most common type of waste.
- Gas diffusion layers that are designed to resist oxidization.

A typical fuel cell produces a voltage from 0.6 V to 0.7 V at full rated load. Voltage decreases as current increases, due to several factors:

- Activation loss.
- Ohmic loss (voltage drop due to resistance of the cell components and interconnections).
- Mass transport loss (depletion of reactants at catalyst sites under high loads, causing rapid loss of voltage).

To deliver the desired amount of energy, the fuel cells can be combined in series to yield higher voltage, and in parallel to allow a higher current to be supplied. Such a design is called a fuel cell stack. The cell surface area can also be increased, to allow higher current from each cell. Within the stack, reactant gases must be distributed uniformly over each of the cells to maximize the power output.

Proton Exchange Membrane Fuel Cells (Pemfcs)

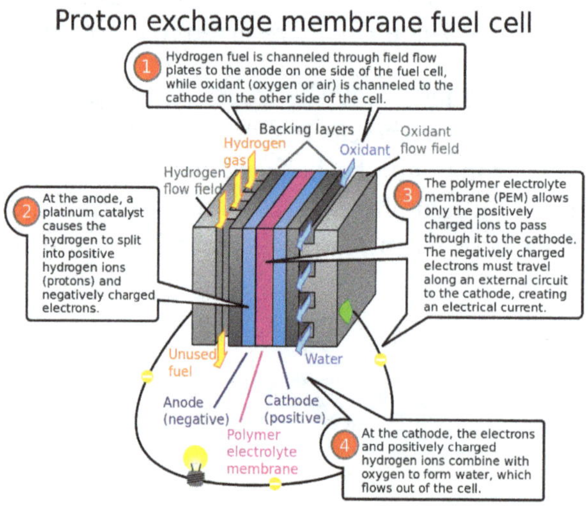

Construction of a high-temperature PEMFC: Bipolar plate as electrode with in-milled gas channel structure, fabricated from conductive composites (enhanced with graphite, carbon black, carbon fiber, and carbon nanotubes for more conductivity); Porous carbon papers; reactive layer, usually on the polymer membrane applied; polymer membrane.

Condensation of water produced by a PEMFC on the air channel wall. The gold wire around the cell ensures the collection of electric current.

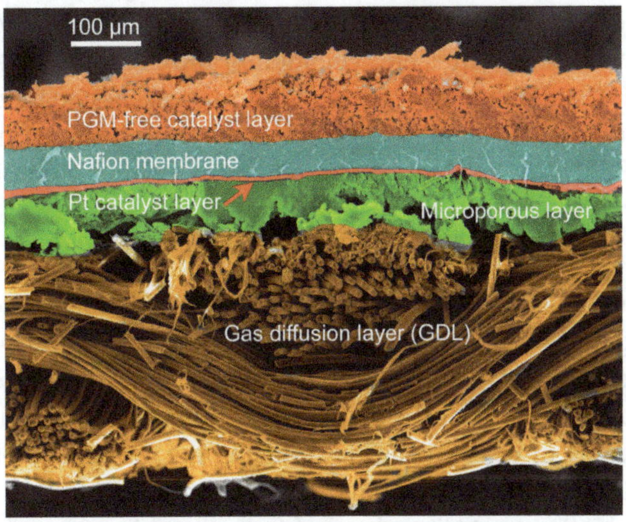

SEM micrograph of a PEMFC MEA cross-section with a non-precious metal catalyst cathode and Pt/C anode. False colors applied for clarity.

In the archetypical hydrogen–oxide proton exchange membrane fuel cell design, a proton-conducting polymer membrane (typically nafion) contains the electrolyte solution that separates the anode and cathode sides. This was called a "solid polymer electrolyte fuel cell" (SPEFC) in the early 1970s, before the proton exchange mechanism was well understood. (Notice that the synonyms "polymer electrolyte membrane" and "proton exchange mechanism" result in the same acronym).

On the anode side, hydrogen diffuses to the anode catalyst where it later dissociates into protons and electrons. These protons often react with oxidants causing them to become what are commonly referred to as multi-facilitated proton membranes. The protons are conducted through the membrane to the cathode, but the electrons are forced to travel in an external circuit (supplying power) because the membrane is electrically insulating. On the cathode catalyst, oxygen molecules react with the electrons (which have traveled through the external circuit) and protons to form water.

In addition to this pure hydrogen type, there are hydrocarbon fuels for fuel cells, including diesel, methanol and chemical hydrides. The waste products with these types of fuel are carbon dioxide and water. When hydrogen is used, the CO2 is released when methane from natural gas is combined with steam, in a process called steam methane reforming, to produce the hydrogen. This can take place in a different location to the fuel cell, potentially allowing the hydrogen fuel cell to be used indoors—for example, in fork lifts.

The different components of a PEMFC are:

- Bipolar plates,
- Electrodes,
- Catalyst,
- Membrane,
- The necessary hardware such as current collectors and gaskets.

The materials used for different parts of the fuel cells differ by type. The bipolar plates may be made of different types of materials, such as, metal, coated metal, graphite, flexible graphite, C–C composite, carbon–polymer composites etc. The membrane electrode assembly (MEA) is referred as the heart of the PEMFC and is usually made of a proton exchange membrane sandwiched between two catalyst-coated carbon papers. Platinum and similar type of noble metals are usually used as the catalyst for PEMFC. The electrolyte could be a polymer membrane.

Proton Exchange Membrane Fuel Cell Design Issues

- Cost. In 2013, the Department of Energy estimated that 80-kW automotive fuel cell system costs of US$67 per kilowatt could be achieved, assuming volume production of 100,000 automotive units per year and US$55 per kilowatt could be achieved, assuming volume production of 500,000 units per year. Many companies are working on techniques to reduce cost in a variety of ways including reducing the amount of platinum needed in each individual cell. Ballard Power Systems has experimented with a catalyst enhanced with carbon silk, which allows a 30% reduction (1 mg/cm^2 to 0.7 mg/cm^2) in platinum usage without reduction in performance. Monash University, Melbourne uses PEDOT as a cathode. A 2011 published study doi: 10.1021/ja1112904 documented the first metal-free electrocatalyst using relatively inexpensive doped carbon nanotubes, which are less than 1% the cost of platinum and are of equal or superior performance. A recently published article demonstrated how the environmental burdens change when using carbon nanotubes as carbon substrate for platinum.
- Water and air management (in PEMFCs). In this type of fuel cell, the membrane must be hydrated, requiring water to be evaporated at precisely the same rate that it is produced. If water is evaporated too quickly, the membrane dries, resistance across it increases, and eventually it will crack, creating a gas "short circuit" where hydrogen and oxygen combine directly, generating heat that will damage the fuel cell. If the water is evaporated too slowly, the electrodes will flood, preventing the reactants from reaching the catalyst and stopping the reaction. Methods to manage water in cells are being developed like electroosmotic pumps focusing on flow control. Just as in a combustion engine, a steady ratio between the reactant and oxygen is necessary to keep the fuel cell operating efficiently.
- Temperature management. The same temperature must be maintained throughout the cell in order to prevent destruction of the cell through thermal loading. This is particularly challenging as the $2H_2 + O_2 \rightarrow 2H_2O$ reaction is highly exothermic, so a large quantity of heat is generated within the fuel cell.

- Durability, service life, and special requirements for some type of cells. Stationary fuel cell applications typically require more than 40,000 hours of reliable operation at a temperature of −35° C to 40° C (−31° F to 104° F), while automotive fuel cells require a 5,000-hour lifespan (the equivalent of 240,000 km (150,000 mi)) under extreme temperatures. Current service life is 2,500 hours (about 75,000 miles). Automotive engines must also be able to start reliably at −30° C (−22° F) and have a high power-to-volume ratio (typically 2.5 kW per liter).
- Limited carbon monoxide tolerance of some (non-PEDOT) cathodes.

Phosphoric Acid Fuel Cell (Pafc)

Phosphoric acid fuel cells (PAFC) were first designed and introduced in 1961 by G. V. Elmore and H. A. Tanner. In these cells phosphoric acid is used as a non-conductive electrolyte to pass positive hydrogen ions from the anode to the cathode. These cells commonly work in temperatures of 150 to 200 degrees Celsius. This high temperature will cause heat and energy loss if the heat is not removed and used properly. This heat can be used to produce steam for air conditioning systems or any other thermal energy consuming system. Using this heat in cogeneration can enhance the efficiency of phosphoric acid fuel cells from 40–50% to about 80%. Phosphoric acid, the electrolyte used in PAFCs, is a non-conductive liquid acid which forces electrons to travel from anode to cathode through an external electrical circuit. Since the hydrogen ion production rate on the anode is small, platinum is used as catalyst to increase this ionization rate. A key disadvantage of these cells is the use of an acidic electrolyte. This increases the corrosion or oxidation of components exposed to phosphoric acid.

Solid Acid Fuel Cell (Safc)

Solid acid fuel cells (SAFCs) are characterized by the use of a solid acid material as the electrolyte. At low temperatures, solid acids have an ordered molecular structure like most salts. At warmer temperatures (between 140 and 150 degrees Celsius for $CsHSO_4$), some solid acids undergo a phase transition to become highly disordered "superprotonic" structures, which increases conductivity by several orders of magnitude. The first proof-of-concept SAFCs were developed in 2000 using cesium hydrogen sulfate ($CsHSO_4$). Current SAFC systems use cesium dihydrogen phosphate (CsH_2PO_4) and have demonstrated lifetimes in the thousands of hours.

Alkaline Fuel Cell (Afc)

The alkaline fuel cell or hydrogen-oxygen fuel cell was designed and first demonstrated publicly by Francis Thomas Bacon in 1959. It was used as a primary source of electrical energy in the Apollo space program. The cell consists of two porous carbon electrodes impregnated with a suitable catalyst such as Pt, Ag, CoO, etc. The space between the two electrodes is filled with a concentrated solution of KOH or NaOH which serves as an electrolyte. H2 gas and O2 gas are bubbled into the electrolyte through the porous carbon electrodes. Thus the overall reaction involves the combination of hydrogen gas and oxygen gas to form water. The cell runs continuously until the reactant's supply is exhausted. This type of cell operates efficiently in the temperature range 343 K to 413 K and provides a potential of about 0.9 V. AAEMFC is a type of AFC which employs a solid polymer electrolyte instead of aqueous potassium hydroxide (KOH) and it is superior to aqueous AFC.

High-temperature Fuel Cells

SOFC

Solid oxide fuel cells (SOFCs) use a solid material, most commonly a ceramic material called yttria-stabilized zirconia (YSZ), as the electrolyte. Because SOFCs are made entirely of solid materials, they are not limited to the flat plane configuration of other types of fuel cells and are often designed as rolled tubes. They require high operating temperatures (800–1000° C) and can be run on a variety of fuels including natural gas.

SOFCs are unique since in those, negatively charged oxygen ions travel from the cathode (positive side of the fuel cell) to the anode (negative side of the fuel cell) instead of positively charged hydrogen ions travelling from the anode to the cathode, as is the case in all other types of fuel cells. Oxygen gas is fed through the cathode, where it absorbs electrons to create oxygen ions. The oxygen ions then travel through the electrolyte to react with hydrogen gas at the anode. The reaction at the anode produces electricity and water as by-products. Carbon dioxide may also be a by-product depending on the fuel, but the carbon emissions from an SOFC system are less than those from a fossil fuel combustion plant. The chemical reactions for the SOFC system can be expressed as follows:

Anode Reaction: $2H_2 + 2O_2^- \rightarrow 2H_2O + 4e^-$

Cathode Reaction: $O_2 + 4e- \rightarrow 2O_2^-$

Overall Cell Reaction: $2H_2 + O_2 \rightarrow 2H_2O$

SOFC systems can run on fuels other than pure hydrogen gas. However, since hydrogen is necessary for the reactions listed above, the fuel selected must contain hydrogen atoms. For the fuel cell to operate, the fuel must be converted into pure hydrogen gas. SOFCs are capable of internally reforming light hydrocarbons such as methane (natural gas), propane and butane. These fuel cells are at an early stage of development.

Challenges exist in SOFC systems due to their high operating temperatures. One such challenge is the potential for carbon dust to build up on the anode, which slows down the internal reforming process. Research to address this "carbon coking" issue at the University of Pennsylvania has shown that the use of copper-based cermet (heat-resistant materials made of ceramic and metal) can reduce coking and the loss of performance. Another disadvantage of SOFC systems is slow start-up time, making SOFCs less useful for mobile applications. Despite these disadvantages, a high operating temperature provides an advantage by removing the need for a precious metal catalyst like platinum, thereby reducing cost. Additionally, waste heat from SOFC systems may be captured and reused, increasing the theoretical overall efficiency to as high as 80%–85%.

The high operating temperature is largely due to the physical properties of the YSZ electrolyte. As temperature decreases, so does the ionic conductivity of YSZ. Therefore, to obtain optimum performance of the fuel cell, a high operating temperature is required. According to their website, Ceres Power, a UK SOFC fuel cell manufacturer, has developed a method of reducing the operating temperature of their SOFC system to 500–600 degrees Celsius. They replaced the commonly used YSZ electrolyte with a CGO (cerium gadolinium oxide) electrolyte. The lower operating temperature allows them to use stainless steel instead of ceramic as the cell substrate, which reduces cost and start-up time of the system.

MCFC

Molten carbonate fuel cells (MCFCs) require a high operating temperature, 650° C (1,200° F), similar to SOFCs. MCFCs use lithium potassium carbonate salt as an electrolyte, and this salt liquefies at high temperatures, allowing for the movement of charge within the cell – in this case, negative carbonate ions.

Like SOFCs, MCFCs are capable of converting fossil fuel to a hydrogen-rich gas in the anode, eliminating the need to produce hydrogen externally. The reforming process creates CO_2 emissions. MCFC-compatible fuels include natural gas, biogas and gas produced from coal. The hydrogen in the gas reacts with carbonate ions from the electrolyte to produce water, carbon dioxide, electrons and small amounts of other chemicals. The electrons travel through an external circuit creating electricity and return to the cathode. There, oxygen from the air and carbon dioxide recycled from the anode react with the electrons to form carbonate ions that replenish the electrolyte, completing the circuit. The chemical reactions for an MCFC system can be expressed as follows:

Anode Reaction: $CO_3^{2-} + H_2 \rightarrow H_2O + CO_2 + 2e^-$

Cathode Reaction: $CO_2 + \frac{1}{2}O_2 + 2e^- \rightarrow CO_3^{2-}$

Overall Cell Reaction: $H_2 + \frac{1}{2}O_2 \rightarrow H_2O$

As with SOFCs, MCFC disadvantages include slow start-up times because of their high operating temperature. This makes MCFC systems not suitable for mobile applications, and this technology will most likely be used for stationary fuel cell purposes. The main challenge of MCFC technology is the cells' short life span. The high-temperature and carbonate electrolyte lead to corrosion of the anode and cathode. These factors accelerate the degradation of MCFC components, decreasing the durability and cell life. Researchers are addressing this problem by exploring corrosion-resistant materials for components as well as fuel cell designs that may increase cell life without decreasing performance.

MCFCs hold several advantages over other fuel cell technologies, including their resistance to impurities. They are not prone to "carbon coking", which refers to carbon build-up on the anode that results in reduced performance by slowing down the internal fuel reforming process. Therefore, carbon-rich fuels like gases made from coal are compatible with the system. The Department of Energy claims that coal, itself, might even be a fuel option in the future, assuming the system can be made resistant to impurities such as sulfur and particulates that result from converting coal into hydrogen. MCFCs also have relatively high efficiencies. They can reach a fuel-to-electricity efficiency of 50%, considerably higher than the 37–42% efficiency of a phosphoric acid fuel cell plant. Efficiencies can be as high as 65% when the fuel cell is paired with a turbine, and 85% if heat is captured and used in a Combined Heat and Power (CHP) system.

Fuel Cell Energy, a Connecticut-based fuel cell manufacturer, develops and sells MCFC fuel cells. The company says that their MCFC products range from 300 kW to 2.8 MW systems that achieve 47% electrical efficiency and can utilize CHP technology to obtain higher overall efficiencies. One product, the DFC-ERG, is combined with a gas turbine and, according to the company, it achieves an electrical efficiency of 65%.

Electric Storage Fuel Cell

The electric storage fuel cell is a conventional battery chargeable by electric power input, using the conventional electro-chemical effect. However, the battery further includes hydrogen (and oxygen) inputs for alternatively charging the battery chemically.

Efficiency of Leading Fuel Cell Types

Theoretical Maximum Efficiency

The energy efficiency of a system or device that converts energy is measured by the ratio of the amount of useful energy put out by the system ("output energy") to the total amount of energy that is put in ("input energy") or by useful output energy as a percentage of the total input energy. In the case of fuel cells, useful output energy is measured in electrical energy produced by the system. Input energy is the energy stored in the fuel. According to the U.S. Department of Energy, fuel cells are generally between 40–60% energy efficient. This is higher than some other systems for energy generation. For example, the typical internal combustion engine of a car is about 25% energy efficient. In combined heat and power (CHP) systems, the heat produced by the fuel cell is captured and put to use, increasing the efficiency of the system to up to 85–90%.

The theoretical maximum efficiency of any type of power generation system is never reached in practice, and it does not consider other steps in power generation, such as production, transportation and storage of fuel and conversion of the electricity into mechanical power. However, this calculation allows the comparison of different types of power generation. The maximum theoretical energy efficiency of a fuel cell is 83%, operating at low power density and using pure hydrogen and oxygen as reactants (assuming no heat recapture) According to the World Energy Council, this compares with a maximum theoretical efficiency of 58% for internal combustion engines.

In Practice

In a fuel-cell vehicle the tank-to-wheel efficiency is greater than 45% at low loads and shows average values of about 36% when a driving cycle like the NEDC (New European Driving Cycle) is used as test procedure. The comparable NEDC value for a Diesel vehicle is 22%. In 2008 Honda released a demonstration fuel cell electric vehicle (the Honda FCX Clarity) with fuel stack claiming a 60% tank-to-wheel efficiency.

It is also important to take losses due to fuel production, transportation, and storage into account. Fuel cell vehicles running on compressed hydrogen may have a power-plant-to-wheel efficiency of 22% if the hydrogen is stored as high-pressure gas, and 17% if it is stored as liquid hydrogen. Fuel cells cannot store energy like a battery, except as hydrogen, but in some applications, such as stand-alone power plants based on discontinuous sources such as solar or wind power, they are combined with electrolyzers and storage systems to form an energy storage system. Most hydrogen is used for oil refining, chemicals and fertilizer production and therefore produced by steam methane reforming, which emits carbon dioxide. The overall efficiency (electricity to hydrogen and back to electricity) of such plants (known as round-trip efficiency), using pure hydrogen and pure oxygen can be "from 35 up to 50 percent", depending on gas density and other conditions.. The electrolyzer/fuel cell system can store indefinite quantities of hydrogen, and is therefore suited for long-term storage.

Solid-oxide fuel cells produce heat from the recombination of the oxygen and hydrogen. The ceramic can run as hot as 800 degrees Celsius. This heat can be captured and used to heat water in a micro combined heat and power (m-CHP) application. When the heat is captured, total efficiency can reach 80–90% at the unit, but does not consider production and distribution losses. CHP units are being developed today for the European home market.

Professor Jeremy P. Meyers, in the Electrochemical Society journal Interface in 2008, wrote, "While fuel cells are efficient relative to combustion engines, they are not as efficient as batteries, due primarily to the inefficiency of the oxygen reduction reaction (and the oxygen evolution reaction, should the hydrogen be formed by electrolysis of water). They make the most sense for operation disconnected from the grid, or when fuel can be provided continuously. For applications that require frequent and relatively rapid start-ups where zero emissions are a requirement, as in enclosed spaces such as warehouses, and where hydrogen is considered an acceptable reactant, a [PEM fuel cell] is becoming an increasingly attractive choice [if exchanging batteries is inconvenient]". In 2013 military organizations are evaluating fuel cells to significantly reduce the battery weight carried by soldiers.

Applications

Type 212 submarine with fuel cell propulsion

Power

Stationary fuel cells are used for commercial, industrial and residential primary and backup power generation. Fuel cells are very useful as power sources in remote locations, such as spacecraft, remote weather stations, large parks, communications centers, rural locations including research stations, and in certain military applications. A fuel cell system running on hydrogen can be compact and lightweight, and have no major moving parts. Because fuel cells have no moving parts and do not involve combustion, in ideal conditions they can achieve up to 99.9999% reliability. This equates to less than one minute of downtime in a six-year period.

Since fuel cell electrolyzer systems do not store fuel in themselves, but rather rely on external storage units, they can be successfully applied in large-scale energy storage, rural areas being one example. There are many different types of stationary fuel cells so efficiencies vary, but most are between 40% and 60% energy efficient. However, when the fuel cell's waste heat is used to heat a building in a cogeneration system this efficiency can increase to 85%. This is significantly more efficient than traditional coal power plants, which are only about one third energy efficient. Assuming production at scale, fuel cells could save 20–40% on energy costs when used in cogeneration systems. Fuel cells are also much cleaner than traditional power generation; a fuel cell power plant using natural gas as a hydrogen source would create less than one ounce of pollution (other than CO2) for every 1,000 kW·h produced, compared to 25 pounds of pollutants generated by conventional combustion systems. Fuel Cells also produce 97% less nitrogen oxide emissions than conventional coal-fired power plants.

One such pilot program is operating on Stuart Island in Washington State. There the Stuart Island Energy Initiative has built a complete, closed-loop system: Solar panels power an electrolyzer, which makes hydrogen. The hydrogen is stored in a 500-U.S.-gallon (1,900 L) tank at 200 pounds per square inch (1,400 kPa), and runs a ReliOn fuel cell to provide full electric back-up to the off-the-grid residence. Another closed system loop was unveiled in late 2011 in Hempstead, NY.

Fuel cells can be used with low-quality gas from landfills or waste-water treatment plants to generate power and lower methane emissions. A 2.8 MW fuel cell plant in California is said to be the largest of the type.

Cogeneration

Combined heat and power (CHP) fuel cell systems, including Micro combined heat and power (MicroCHP) systems are used to generate both electricity and heat for homes, office building and factories. The system generates constant electric power (selling excess power back to the grid when it is not consumed), and at the same time produces hot air and water from the waste heat. As the result CHP systems have the potential to save primary energy as they can make use of waste heat which is generally rejected by thermal energy conversion systems. A typical capacity range of home fuel cell is 1–3 kWel/4–8 kWth. CHP systems linked to absorption chillers use their waste heat for refrigeration.

The waste heat from fuel cells can be diverted during the summer directly into the ground providing further cooling while the waste heat during winter can be pumped directly into the building. The University of Minnesota owns the patent rights to this type of system

Co-generation systems can reach 85% efficiency (40–60% electric + remainder as thermal). Phosphoric-acid fuel cells (PAFC) comprise the largest segment of existing CHP products worldwide and can provide combined efficiencies close to 90%. Molten Carbonate (MCFC) and Solid Oxide Fuel Cells (SOFC) are also used for combined heat and power generation and have electrical energy efficiences around 60%. Disadvantages of co-generation systems include slow ramping up and down rates, high cost and short lifetime. Also their need to have a hot water storage tank to smooth out the thermal heat production was a serious disadvantage in the domestic market place where space in domestic properties is at a great premium.

Delta-ee consultants stated in 2013 that with 64% of global sales the fuel cell micro-combined heat and power passed the conventional systems in sales in 2012. The Japanese ENE FARM project will pass 100,000 FC mCHP systems in 2014, 34.213 PEMFC and 2.224 SOFC were installed in the period 2012-2014, 30,000 units on LNG and 6,000 on LPG.

Fuel Cell Electric Vehicles (Fcevs)

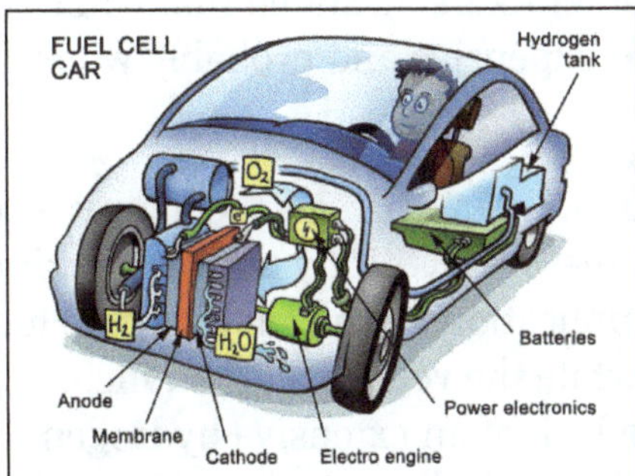

Configuration of components in a fuel cell car

Toyota Mirai

Element One fuel cell vehicle

Automobiles

As of 2017, about 6500 FCEVs have been leased or sold worldwide. Three fuel cell electric vehicles have been introduced for commercial lease and sale: the Honda Clarity, Toyota Mirai and the Hyundai ix35 FCEV. Additional demonstration models include the Honda FCX Clarity, and Mercedes-Benz F-Cell. As of June 2011 demonstration FCEVs had driven more than 4,800,000 km (3,000,000 mi), with more than 27,000 refuelings. Fuel cell electric vehicles feature an average range of 314 miles between refuelings. They can be refueled in less than 5 minutes. The U.S. Department of Energy's Fuel Cell Technology Program states that, as of 2011, fuel cells achieved 53–59% efficiency at one-quarter power and 42–53% vehicle efficiency at full power, and a durability of over 120,000 km (75,000 mi) with less than 10% degradation. In a Well-to-Wheels simulation analysis that "did not address the economics and market constraints", General Motors and its partners estimated that per mile traveled, a fuel cell electric vehicle running on compressed gaseous hydrogen produced from natural gas could use about 40% less energy and emit 45% less greenhouse gasses than an internal combustion vehicle. A lead engineer from the Department of Energy whose team is testing fuel cell cars said in 2011 that the potential appeal is that "these are full-function vehicles with no limitations on range or refueling rate so they are a direct replacement for any vehicle. For instance, if you drive a full sized SUV and pull a boat up into the mountains, you can do that with this technology and you can't with current battery-only vehicles, which are more geared toward city driving."

In 2015, Toyota introduced its first fuel cell vehicle, the Mirai, at a price of $57,000. Hyundai introduced the limited production Hyundai ix35 FCEV under a lease agreement.. In 2016, Honda started leasing the Honda Clarity Fuel Cell.

Criticism

Some experts believe that hydrogen fuel cell cars will never become economically competitive with other technologies or that it will take decades for them to become profitable. Elon Musk stated in 2015 that fuel cells for use in cars will never be commercially viable because of the inefficiency of producing, transporting and storing hydrogen and the flammability of the gas, among other reasons. Professor Jeremy P. Meyers estimated in 2008 that cost reductions over a production ramp-up period will take about 20 years after fuel-cell cars are introduced before they will be able to compete commercially with current market technologies, including gasoline internal combustion engines. In 2011, the chairman and CEO of General Motors, Daniel Akerson, stated that while the cost of hydrogen fuel cell cars is decreasing: "The car is still too expensive and probably won't be practical until the 2020-plus period, I don't know."

In 2012, Lux Research, Inc. issued a report that stated: "The dream of a hydrogen economy is no nearer". It concluded that "Capital cost will limit adoption to a mere 5.9 GW" by 2030, providing "a nearly insurmountable barrier to adoption, except in niche applications". The analysis concluded that, by 2030, PEM stationary market will reach $1 billion, while the vehicle market, including forklifts, will reach a total of $2 billion. Other analyses cite the lack of an extensive hydrogen infrastructure in the U.S. as an ongoing challenge to Fuel Cell Electric Vehicle commercialization. In 2006, a study for the IEEE showed that for hydrogen produced via electrolysis of water: "Only about 25% of the power generated from wind, water, or sun is converted to practical use." The study further noted that "Electricity obtained from hydrogen fuel cells appears to be four times as expensive as electricity drawn from the electrical transmission grid. Because of the high energy losses [hydrogen] cannot compete with electricity." Furthermore, the study found: "Natural gas reforming is not a sustainable solution". "The large amount of energy required to isolate hydrogen from natural compounds (water, natural gas, biomass), package the light gas by compression or liquefaction, transfer the energy carrier to the user, plus the energy lost when it is converted to useful electricity with fuel cells, leaves around 25% for practical use."

In 2014, Joseph Romm, the author of The Hype About Hydrogen (2005), stated that FCVs still had not overcome the high fueling cost, lack of fuel-delivery infrastructure, and pollution caused by producing hydrogen. "It would take several miracles to overcome all of those problems simultaneously in the coming decades." He concluded that renewable energy cannot economically be used to make hydrogen for an FCV fleet "either now or in the future." Greentech Media's analyst reached similar conclusions in 2014. In 2015, Clean Technica listed some of the disadvantages of hydrogen fuel cell vehicles. So did Car Throttle.

Buses

Toyota FCHV-BUS at the Expo 2005.

As of August 2011, there were a total of approximately 100 fuel cell buses deployed around the world. Most buses are produced by UTC Power, Toyota, Ballard, Hydrogenics, and Proton Motor. UTC buses had accumulated over 970,000 km (600,000 mi) of driving by 2011. Fuel cell buses have a 39–141% higher fuel economy than diesel buses and natural gas buses. Fuel cell buses have been deployed around the world including in Whistler, Canada; San Francisco, United States; Hamburg, Germany; Shanghai, China; London, England; and São Paulo, Brazil.

The Fuel Cell Bus Club is a global cooperative effort in trial fuel cell buses. Notable projects include:

- 12 fuel cell buses are being deployed in the Oakland and San Francisco Bay area of California.
- Daimler AG, with 36 experimental buses powered by Ballard Power Systems fuel cells, completed a successful three-year trial in eleven cities, in January 2007.
- A fleet of Thor buses with UTC Power fuel cells was deployed in California, operated by SunLine Transit Agency.

The first Brazilian hydrogen fuel cell bus prototype in Brazil was deployed in São Paulo. The bus was manufactured in Caxias do Sul and the hydrogen fuel will be produced in São Bernardo do Campo from water through electrolysis.

Forklifts

A fuel cell forklift (also called a fuel cell lift truck) is a fuel cell-powered industrial forklift truck used to lift and transport materials. In 2013 there were over 4,000 fuel cell forklifts used in material handling in the US, of which only 500 received funding from DOE (2012). The global market is 1 million fork lifts per year. Fuel cell fleets are operated by various companies, including Sysco Foods, FedEx Freight, GENCO (at Wegmans, Coca-Cola, Kimberly Clark, and Whole Foods), and H-E-B Grocers. Europe demonstrated 30 fuel cell forklifts with Hylift and extended it with HyLIFT-EUROPE to 200 units, with other projects in France and Austria. Pike Research stated in 2011 that fuel cell-powered forklifts will be the largest driver of hydrogen fuel demand by 2020.

Most companies in Europe and the US do not use petroleum-powered forklifts, as these vehicles work indoors where emissions must be controlled and instead use electric forklifts. Fuel cell-powered forklifts can provide benefits over battery-powered forklifts as they can work for a full 8-hour shift on a single tank of hydrogen and can be refueled in 3 minutes. Fuel cell-powered forklifts can be used in refrigerated warehouses, as their performance is not degraded by lower temperatures. The FC units are often designed as drop-in replacements.

Motorcycles and Bicycles

In 2005 a British manufacturer of hydrogen-powered fuel cells, Intelligent Energy (IE), produced the first working hydrogen run motorcycle called the ENV (Emission Neutral Vehicle). The motorcycle holds enough fuel to run for four hours, and to travel 160 km (100 mi) in an urban area, at a top speed of 80 km/h (50 mph). In 2004 Honda developed a fuel-cell motorcycle that utilized the Honda FC Stack.

Other examples of motorbikes and bicycles that use hydrogen fuel cells include the Taiwanese company APFCT's scooter using the fueling system from Italy's Acta SpA and the Suzuki Burgman scooter with an IE fuel cell that received EU Whole Vehicle Type Approval in 2011. Suzuki Motor Corp. and IE have announced a joint venture to accelerate the commercialization of zero-emission vehicles.

Airplanes

In 2003, the world's first propeller-driven airplane to be powered entirely by a fuel cell was flown. The fuel cell was a stack design that allowed the fuel cell to be integrated with the plane's aerodynamic surfaces. Fuel cell-powered unmanned aerial vehicles (UAV) include a Horizon fuel cell UAV that set the record distance flown for a small UAV in 2007. Boeing researchers and industry partners throughout Europe conducted experimental flight tests in February 2008 of a manned airplane powered only by a fuel cell and lightweight batteries. The fuel cell demonstrator airplane, as it was called, used a proton exchange membrane (PEM) fuel cell/lithium-ion battery hybrid system to power an electric motor, which was coupled to a conventional propeller.

In 2009 the Naval Research Laboratory's (NRL's) Ion Tiger utilized a hydrogen-powered fuel cell and flew for 23 hours and 17 minutes. Fuel cells are also being tested and considered to provide auxiliary power in aircraft, replacing fossil fuel generators that were previously used to start the engines and power on board electrical needs, while reducing carbon emissions.

In 2016 a Raptor E1 drone made a successful test flight using a fuel cell that was lighter than the lithium-ion battery it replaced. The flight lasted 10 minutes at an altitude of 80 metres (260 ft), although the fuel cell reportedly had enough fuel to fly for two hours. The fuel was contained in approximately 100 solid 1 square centimetre (0.16 sq in) pellets composed of a proprietary chemical within an unpressurized cartridge. The pellets are physically robust and operate at temperatures as warm as 50° C (122° F). The cell was from Arcola Energy.

Boats

The world's first certified fuel cell boat (HYDRA), in Leipzig/Germany

The world's first fuel-cell boat HYDRA used an AFC system with 6.5 kW net output. Iceland has committed to converting its vast fishing fleet to use fuel cells to provide auxiliary power by 2015 and, eventually, to provide primary power in its boats. Amsterdam recently introduced its first fuel cell-powered boat that ferries people around the city's canals.

Submarines

The Type 212 submarines of the German and Italian navies use fuel cells to remain submerged for weeks without the need to surface.

The U212A is a non-nuclear submarine developed by German naval shipyard Howaldtswerke Deutsche Werft. The system consists of nine PEM fuel cells, providing between 30 kW and 50 kW each. The ship is silent, giving it an advantage in the detection of other submarines. A naval paper has theorized about the possibility of a nuclear-fuel cell hybrid whereby the fuel cell is used when silent operations are required and then replenished from the Nuclear reactor (and water).

Portable Power Systems

Portable fuel cell systems are generally classified as weighing under 10 kg and providing power of less than 5 kW. The potential market size for smaller fuel cells is quite large with an up to 40% per annum potential growth rate and a market size of around $10 billion, leading a great deal of research to be devoted to the development of portable power cells. Within this market two groups have been identified. The first is the microfuel cell market, in the 1-50 W range for power smaller electronic devices. The second is the 1-5 kW range of generators for larger scale power generation (e.g. military outposts, remote oil fields).

Microfuel cells are primarily aimed at penetrating the market for phones and laptops. This can be primarily attributed to the advantageous energy density provided by fuel cells over a lithium-ion battery, for the entire system. For a battery, this system includes the charger as well as the battery itself. For the fuel cell this system would include the cell, the necessary fuel and peripheral attachments. Taking the full system into consideration, fuel cells have been shown to provide 530Wh/kg compared to 44 Wh/kg for lithium ion batteries. However, while the weight of fuel cell systems offer a distinct advantage the current costs are not in their favor. while a battery system will generally cost around $1.20 per Wh, fuel cell systems cost around $5 per Wh, putting them at a significant disadvantage.

As power demands for cell phones increase, fuel cells could become much more attractive options for larger power generation. The demand for longer on time on phones and computers is something often demanded by consumers so fuel cells could start to make strides into laptop and cell phone markets. The price will continue to go down as developments in fuel cells continues to accelerate. Current strategies for improving micro fuelcells is through the use of carbon nanotubes. It was shown by Girishkumar et al. that depositing nanotubes on electrode surfaces allows for substantially greater surface area increasing the oxygen reduction rate.

Fuel cells for use in larger scale operations also show much promise. Portable power systems that use fuel cells can be used in the leisure sector (i.e. RVs, cabins, marine), the industrial sector (i.e. power for remote locations including gas/oil wellsites, communication towers, security, weather stations), and in the military sector. SFC Energy is a German manufacturer of direct methanol fuel cells for a variety of portable power systems. Ensol Systems Inc. is an integrator of portable power systems, using the SFC Energy DMFC. The key advantage of fuel cells in this market is the great power generation per weight. While fuel cells can be expensive, for remote locations that require dependable energy fuel cells hold great power. For a 72-h excursion the comparison in weight is substantial, with a fuel cell only weighing 15 pounds compared to 29 pounds of batteries needed for the same energy.

Other Applications

- Providing power for base stations or cell sites.
- Distributed generation.
- Emergency power systems are a type of fuel cell system, which may include lighting, generators and other apparatus, to provide backup resources in a crisis or when regular systems fail. They find uses in a wide variety of settings from residential homes to hospitals, scientific laboratories, data centers, telecommunication equipment and modern naval ships.
- An uninterrupted power supply (UPS) provides emergency power and, depending on the topology, provide line regulation as well to connected equipment by supplying power from a separate source when utility power is not available. Unlike a standby generator, it can provide instant protection from a momentary power interruption.
- Base load power plants.
- Solar Hydrogen Fuel Cell Water Heating.
- Hybrid vehicles, pairing the fuel cell with either an ICE or a battery.
- Notebook computers for applications where AC charging may not be readily available.
- Portable charging docks for small electronics (e.g. a belt clip that charges a cell phone or PDA).
- Smartphones, laptops and tablets.
- Small heating appliances.
- Food preservation, achieved by exhausting the oxygen and automatically maintaining oxygen exhaustion in a shipping container, containing, for example, fresh fish.
- Breathalyzers, where the amount of voltage generated by a fuel cell is used to determine the concentration of fuel (alcohol) in the sample.
- Carbon monoxide detector, electrochemical sensor.

Fueling Stations

In 2013, The New York Times reported that there were "10 hydrogen stations available to the public in the entire United States: one in Columbia, S.C., eight in Southern California and the one in Emeryville". As of December 2016, there were 31 publicly accessible hydrogen refueling stations in the US, 28 of which were located in California.

A public hydrogen refueling station in Iceland operated from 2003 to 2007. It served three buses in the public transport net of Reykjavík. The station produced its own hydrogen with an electrolyzing unit. The 14 stations in Germany were planned to be expanded to 50 by 2015 through its public–private partnership Now GMBH.

Hydrogen fueling station.

By May 2017, there were 91 hydrogen fueling stations in Japan. As of 2016, Norway planned to build a network of hydrogen stations between the major cities, starting in 2017.

Markets and Economics

In 2012, fuel cell industry revenues exceeded $1 billion market value worldwide, with Asian pacific countries shipping more than 3/4 of the fuel cell systems worldwide. However, as of January 2014, no public company in the industry had yet become profitable. There were 140,000 fuel cell stacks shipped globally in 2010, up from 11,000 shipments in 2007, and from 2011 to 2012 worldwide fuel cell shipments had an annual growth rate of 85%. Tanaka Kikinzoku expanded its manufacturing facilities in 2011. Approximately 50% of fuel cell shipments in 2010 were stationary fuel cells, up from about a third in 2009, and the four dominant producers in the Fuel Cell Industry were the United States, Germany, Japan and South Korea. The Department of Energy Solid State Energy Conversion Alliance found that, as of January 2011, stationary fuel cells generated power at approximately $724 to $775 per kilowatt installed. In 2011, Bloom Energy, a major fuel cell supplier, said that its fuel cells generated power at 9–11 cents per kilowatt-hour, including the price of fuel, maintenance, and hardware.

Industry groups predict that there are sufficient platinum resources for future demand, and in 2007, research at Brookhaven National Laboratory suggested that platinum could be replaced by a gold-palladium coating, which may be less susceptible to poisoning and thereby improve fuel cell lifetime. Another method would use iron and sulphur instead of platinum. This would lower the cost of a fuel cell (as the platinum in a regular fuel cell costs around US$1,500, and the same amount of iron costs only around US$1.50). The concept was being developed by a coalition of the John Innes Centre and the University of Milan-Bicocca. PEDOT cathodes are immune to monoxide poisoning.

In 2016, Samsung "decided to drop fuel cell-related business projects, as the outlook of the market isn't good".

The Advantages and Disadvantages of Fuel Cells

If compared with conventional fossil fuel propelled electric generators, the use of fuel cells brings about many advantages:

- Higher volumetric and gravimetric efficiency.
- Low chemical, acoustic, and thermal emissions.
- Modularity and siting flexibility.
- Low maintenance.
- Fuel flexibility (depending on type of fuel cell).
- No production of pollutants.

Higher Efficiency

The higher efficiency of fuel cells is a consequence of the direct (chemical) production of electric energy from the fuel used. As a consequence the technology is not affected by the limitations derived from the Carnot thermic cycle which burdens all the combustion based electric generation systems.

Low Chemical, Acoustic, and Thermal Emissions

Due to higher efficiencies and lower fuel oxidation temperatures, fuel cells emit less carbon dioxide and nitrogen oxides for kilowatt of power generated. Another characteristic is that since fuel cells have no moving parts (except for the ancillary pumps, blowers, and transformers), noise and vibration are negligible.

Modularity and Siting Flexibility

A single fuel cell will produce less than one volt of electrical potential. To produce higher voltages, fuel cells are stacked on top of each other and connected in series. Cell stacks consist of repeating fuel cell units, each comprised of an anode, cathode, electrolyte and a bipolar separator plate. The number of cells in a stack depends on the desired power output and individual cell performance. The stacks have sizes which ranges from a few hundreds of W to several hundred of kW (up to some MW).

Low Maintenance

Due to the high modularity of generator systems it is relatively easy, for same type of fuel cells, to localize and substitute a damaged or malfunctioning cell contained inside a stack. This characteristic obviously leads to lower maintenance costs.

Fuel Flexibility

Hydrogen is the most used fuel (especially for the low temperature technologies of fuel cells which requires pure gasses to operate); however fuel flexibility has been demonstrated in many technologies using natural gas, propane, landfill gas, anaerobic digester gas, military logistic fuels, and coal gas. This flexibility depends for the most on the operative temperature range of the type of fuel cells used (in principle, the higher the temperature the less pure the gas that the fuel cell can use).

Even if they are able to provide many important advantages, all the fuel cells technologies are still in a development phase and are burdened by many problems which makes their use less convenient than the other technologies which are already in use.

The problems are the following:

- Fuel cells costs for stationary electric generation (€/Wh) are still too high and unsuitable for the substitution of the technologies based on fossil fuels.
- The life cycle and the degradation time of many technologies of fuel cells (especially the high temperature technologies which are the best for electric power generation) is still not totally known.
- Hydrogen, which is one the main fuel for fuel cell technologies, is expensive and it does not still exist a network for his production and distribution.
- The use of low temperature fuel cells in the car market is limited by the fact that it is difficult to confine a suitable quantity of hydrogen in small fuel containers and by the fact that hydrogen is a flammable and potentially explosive gas (especially if compressed in small containers).

These are the reasons which prevents fuel cells to substitute many others technologies with lower efficiency and a much higher environmental impact.

The result is that an high effort in research and development is still needed to render this technology approachable, especially regarding the reduction of energy production costs the use of new materials and the improvement of conductivity mechanism in the electrolytes, also because the development of new materials remain in progress.

Fuel Cell Vehicle

Fuel cell vehicles (FCVs) have the potential to significantly reduce our dependence on foreign oil and lower harmful emissions that contribute to climate change. FCVs run on hydrogen gas rather than gasoline and emit no harmful tailpipe emissions. Several challenges must be overcome for them to be competitive with conventional vehicles, but their potential benefits are substantial.

A Look Inside

FCVs look like conventional vehicles but use cutting edge technologies. The heart of the FCV is the fuel cell stack. The stack converts hydrogen gas stored onboard with oxygen from the air into electricity, which powers the vehicle's electric motor. The major components of a typical FCV are illustrated below:

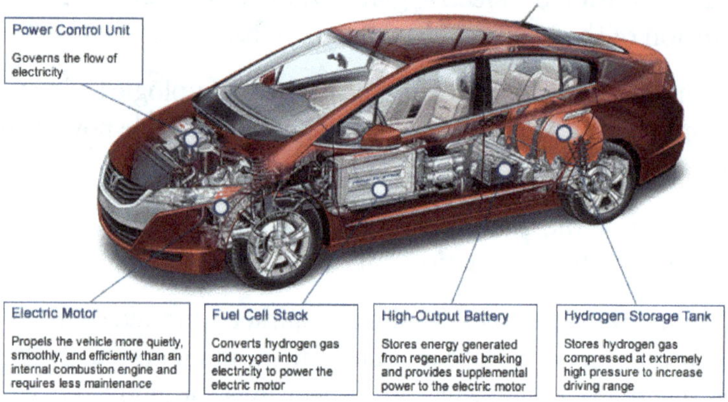

There are fuel cell vehicles for all modes of transport. The most prevalent fuel cell vehicles are cars, buses, forklifts and material handling vehicles.

Automobiles

The Honda FCX Clarity concept car was introduced in 2008 for leasing by customers in Japan and Southern California and discontinued by 2015. From 2008 to 2014, Honda leased a total of 45 FCX units in the US. Over 20 other FCEVs prototypes and demonstration cars were released in that time period, including the GM HydroGen4, and Mercedes-Benz F-Cell.

The Hyundai ix35 FCEV Fuel Cell vehicle has been available for lease since 2014, when 54 units were leased.

Sales of the Toyota Mirai to government and corporate customers began in Japan in December 2014. Pricing started at ¥6,700,000 (~US$57,400) before taxes and a government incentive of ¥2,000,000 (~US$19,600). Former European Parliament President Pat Cox estimated that Toyota initially would lose about $100,000 on each Mirai sold. As of December 2017, global sales totaled 5,300 Mirais. The top selling markets were the U.S. with 2,900 units, Japan with 2,100 and Europe with 200.

Retail deliveries of the 2017 Honda Clarity Fuel Cell began in California in December 2016. The Clarity Fuel Cell, with range of 366 mi (589 km), has the highest EPA driving range rating of any zero-emissions vehicle in the U.S., including fuel cell and battery electric vehicles. The 2017 Clarity also has the highest combined and city fuel economy ratings among all hydrogen fuel cell cars rated by the EPA, with a combined city/highway rating of 67 miles per gallon gasoline equivalent (MPGe), and 68 MPGe in city driving.

In 2017 Daimler phased out of its FCEV development, citing declining battery costs and increasing range of EVs, and most of the automobile companies developing hydrogen cars had switched their focus to battery electric vehicles.

Fuel Economy

The following table compares EPA's fuel economy expressed in miles per gallon gasoline equivalent (MPGe) for the hydrogen fuel cell vehicles rated by the EPA as of December 2016, and available only in California.

Comparison of fuel economy expressed in MPGe for hydrogen fuel cell vehicles available for leasing in California and rated by the U.S. Environmental Protection Agency as of October 2016						
Vehicle	Model year	Combined fuel economy	City fuel economy	Highway fuel economy	Range	Annual fuel cost
Hyundai Tucson Fuel Cell	2017	49 mpg-e	48 mpg-e	50 mpg-e	265 mi (426 km)	US$1,700
Toyota Mirai	2016	66 mpg-e	66 mpg-e	66 mpg-e	312 mi (502 km)	US$1,250
Honda Clarity Fuel Cell	2017	67 mpg-e	68 mpg-e	66 mpg-e	366 mi (589 km)	-
One kg of hydrogen has roughly the same energy content as one U.S. gallon of gasoline.						

Fuel Cells Powered by an Ethanol Reformer

In June 2016, Nissan announced plans to develop fuel cell vehicles powered by ethanol rather than hydrogen. Nissan claims this technical approach would be cheaper, and that it would be easier to deploy the fueling infrastructure than a hydrogen infrastructure. The vehicle would include a tank holding a blend of water and ethanol, which is fed into an onboard reformer that splits it into hydrogen and carbon dioxide. The hydrogen is then fed into a solid oxide fuel cell. According to Nissan, the liquid fuel could be an ethanol-water blend at a 55:45 ratio. Nissan expects to commercialize its technology by 2020.

Buses

Mercedes-Benz fuel cell bus.

There are also demonstration models of buses, and in 2011 there were over 100 fuel cell buses deployed around the world. Most of these buses were produced by UTC Power, Toyota, Ballard,

Hydrogenics, and Proton Motor. UTC buses had accumulated over 970,000 km (600,000 mi) of driving. Fuel cell buses have a 30-141% higher fuel economy than diesel buses and natural gas buses. Fuel cell buses have been deployed in Whistler Canada, San Francisco US, Hamburg Germany, Shanghai China, London England, São Paulo Brazil and several other cities. The Whistler project was discontinued in 2015. The Fuel Cell Bus Club is a global cooperative effort in trial fuel cell buses. Notable Projects Include:

- 12 Fuel cell buses were deployed in the Oakland and San Francisco Bay area of California.
- Daimler AG, with thirty-six experimental buses powered by Ballard Power Systems fuel cells, completed a successful three-year trial, in eleven cities, in 2007.
- A fleet of Thor buses with UTC Power fuel cells was deployed in California, operated by SunLine Transit Agency.
- The first hydrogen fuel cell bus prototype in Brazil was deployed in São Paulo. The bus was manufactured in Caxias do Sul, and the hydrogen fuel was to be produced in São Bernardo do Campo from water through electrolysis. The program, called "Ônibus Brasileiro a Hidrogênio" (Brazilian Hydrogen Autobus), included three buses.

Forklifts

A fuel cell forklift (also called a fuel cell lift truck or a fuel cell forklift) is a fuel cell-powered industrial forklift truck used to lift and transport materials. Most fuel cells used in forklifts are powered by PEM fuel cells.

In 2013 there were over 4,000 fuel cell forklifts used in material handling in the US from which only 500 received funding from DOE (2012). Fuel cell fleets are operated by a large number of companies, including Sysco Foods, FedEx Freight, GENCO (at Wegmans, Coca-Cola, Kimberly Clark, and Whole Foods), and H-E-B Grocers. Europe demonstrated 30 fuel cell forklifts with Hylift and extended it with HyLIFT-EUROPE to 200 units, with other projects in France and Austria. Pike Research stated in 2011 that fuel-cell-powered forklifts will be the largest driver of hydrogen fuel demand by 2020.

PEM fuel-cell-powered forklifts provide significant benefits over petroleum powered forklifts as they produce no local emissions. Fuel-cell forklifts can work for a full 8-hour shift on a single tank of hydrogen, can be refueled in 3 minutes and have a lifetime of 8–10 years. Fuel cell-powered forklifts are often used in refrigerated warehouses as their performance is not degraded by lower temperatures. In design the FC units are often made as drop-in replacements.

Motorcycles and Bicycles

In 2005 the British firm Intelligent Energy produced the first ever working hydrogen run motorcycle called the ENV (Emission Neutral Vehicle). The motorcycle holds enough fuel to run for four hours, and to travel 160 km (100 mi) in an urban area, at a top speed of 80 km/h (50 mph). In 2004 Honda developed a fuel-cell motorcycle which utilized the Honda FC Stack.There are other examples of bikes and bicycles with a hydrogen fuel cell engine. The Suzuki Burgman received "whole vehicle type" approval in the EU. The Taiwanese company APFCT conducts a live street test with 80 fuel cell scooters for Taiwans Bureau of Energy using the fueling system from Italy's Acta SpA.

Yamaha FC-me motorcycle.

Airplanes

The Boeing Fuel Cell Demonstrator powered by a hydrogen fuel cell.

Boeing researchers and industry partners throughout Europe conducted experimental flight tests in February 2008 of a manned airplane powered only by a fuel cell and lightweight batteries. The Fuel Cell Demonstrator Airplane, as it was called, used a Proton Exchange Membrane (PEM) fuel cell/lithium-ion battery hybrid system to power an electric motor, which was coupled to a conventional propeller. In 2003, the world's first propeller driven airplane to be powered entirely by a fuel cell was flown. The fuel cell was a unique FlatStack stack design which allowed the fuel cell to be integrated with the aerodynamic surfaces of the plane.

There have been several fuel cell powered unmanned aerial vehicles (UAV). A Horizon fuel cell UAV set the record distance flow for a small UAV in 2007. The military is especially interested in this application because of the low noise, low thermal signature and ability to attain high altitude. In 2009 the Naval Research Laboratory's (NRL's) Ion Tiger utilized a hydrogen-powered fuel cell and flew for 23 hours and 17 minutes. Boeing is completing tests on the Phantom Eye, a high-altitude, long endurance (HALE) to be used to conduct research and surveillance flying at 20,000 m (65,000 ft) for up to four days at a time. Fuel cells are also being used to provide auxiliary power for aircraft, replacing fossil fuel generators that were previously used to start the engines and power on board electrical needs. Fuel cells can help airplanes reduce CO2 and other pollutant emissions and noise.

Boats

The world's first Fuel Cell Boat HYDRA used an AFC system with 6.5 kW net output. For each liter of fuel consumed, the average outboard motor produces 140 times less the hydrocarbons produced by the average modern car. Fuel cell engines have higher energy efficiencies than combustion engines, and therefore offer better range and significantly reduced emissions. Iceland has committed to converting its vast fishing fleet to use fuel cells to provide auxiliary power by 2015 and, eventually, to provide primary power in its boats. Amsterdam recently introduced its first fuel cell powered boat that ferries people around the city's famous and beautiful canals.

Submarines

The first submersible application of fuel cells is the German Type 212 submarine. Each Type 212 contains nine PEM fuel cells, spread throughout the ship, providing between 30 kW and 50 kW each of electrical power. This allows the Type 212 to remain submerged longer and makes them more difficult to detect. Fuel cell powered submarines are also easier to design, manufacture, and maintain than nuclear-powered submarines.

Trains

Debut of the Alstom Coradia iLint at InnoTrans 2016

In March 2015, China South Rail Corporation (CSR) demonstrated the world's first hydrogen fuel cell-powered tramcar at an assembly facility in Qingdao. The chief engineer of the CSR subsidiary CSR Sifang Co Ltd., Liang Jianying, said that the company is studying how to reduce the running costs of the tram. A total of 83 miles of tracks for the new vehicle have been built in seven Chinese cities. China plans to spend 200 billion yuan ($32 billion) over the next five years to increase tram tracks to more than 1,200 miles.

In 2016, Alstom debuted the Coradia iLint, a regional train powered by hydrogen fuel cells that will be the world's first production hydrogen-powered trainset. The Coradia iLint will be able to reach 140 kilometres per hour (87 mph) and travel 600–800 kilometres (370–500 mi) on a full tank of hydrogen. The first Coradia iLint is expected to enter service in December 2017 on the Buxtehude-Bremervörde-Bremerhaven-Cuxhaven line in Lower Saxony, Germany.

Benefits

Less Greenhouse Gas Emissions

Gasoline and diesel-powered vehicles emit greenhouse gases (GHGs), mostly carbon dioxide (CO_2), that contribute to climate change. Fuel cell vehicles (FCVs) powered by pure hydrogen emit no tailpipe GHGs, only heat and water.

Producing the hydrogen for FCVs can generate GHGs, depending on the production method. Still, it generates much less GHGs than conventional gasoline and diesel vehicles.

Reduced Oil Dependence

FCVs could reduce our dependence on foreign oil since hydrogen can be derived from domestic sources. These sources include natural gas and coal, as well as renewable sources such as water, biogas, and agricultural waste. That would make our economy less dependent on other countries and less vulnerable to oil price shocks from an increasingly volatile oil market.

Less Air Pollutants

Highway vehicles emit much of the air pollutants that contribute to smog and harmful particulates in the U.S. FCVs powered by pure hydrogen emit no harmful pollutants. Producing hydrogen from fossil fuels causes some pollutants, but much less than the amount generated by conventional vehicles.

Challenges

Several challenges must be overcome before FCVs will be a successful, competitive alternative for consumers.

Vehicle Cost

Fuel Cell System Cost Progress

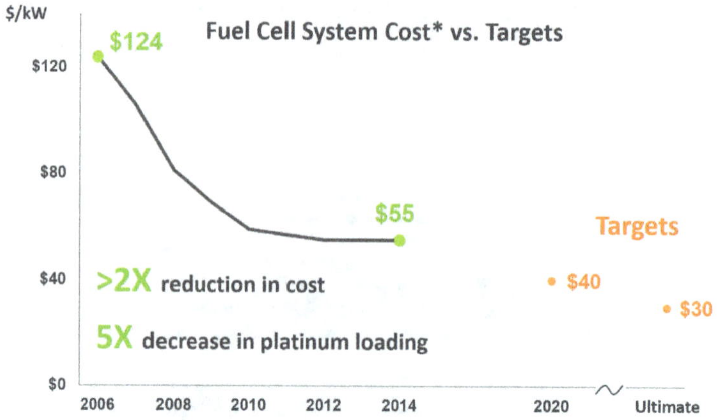

USDOE, Fuel Cell Technologies Office, Fuel Cell Technologies Office Accomplishments and Progress.

FCVs are currently more expensive than conventional vehicles and hybrids. However, costs have decreased significantly and are approaching DOE's goal for 2020. Car makers must continue to lower costs, especially for the fuel cell stack and hydrogen storage, for FCVs to compete with conventional vehicles.

Getting Hydrogen to Consumers

The current infrastructure for producing and getting hydrogen to consumers cannot yet support the widespread adoption of FCVs. In 2013, H2USA was launched as a public-private partnership between DOE and other federal agencies, automakers, state government, academic institutions, and additional stakeholders. Its goal is to coordinate research and identify cost-effective solutions for deploying hydrogen infrastructure. By the end of 2015, more than 50 public stations should be available, mostly in California. This is an important first step in making hydrogen available to consumers.

Fuel Cell Durability and Reliability

Fuel cell systems are not yet as durable as internal combustion engines, especially in some temperature and humidity ranges. On-road fuel cell stack durability is currently about half of what is needed for commercialization. Durability has increased substantially over the past few years from 29,000 miles to 75,000 miles, but experts believe FCVs must achieve a 150,000-mile expected lifetime to compete with gasoline vehicles.

Public Education

FCVs must be embraced by consumers before their benefits can be realized. As with any new vehicle technology, consumers may have concerns about the dependability and safety of these vehicles when they first hit the market. Plus, they must become familiar with a new kind of fuel. Public education can accelerate this process.

Fuel Cell Forklift

Fuel cell forklifts, which utilise hydrogen to generate power, can be rapidly refuelled in as fast as three minutes, allowing operators to get back to work quickly. In multi-shift operations with two or more battery replacements per day, the quick refuelling of hydrogen fuel cells saves time and increases operator efficiency, thus increasing productivity. Furthermore, since lift truck operators can refuel hydrogen themselves, operations can keep business moving and make more efficient use of labour resources. In addition, fuel cell powered forklifts are often used in refrigerated warehouses as their performance is not as affected by temperature as some types of lithium batteries.

Toyota's fuel cell forklift

In early 2017, Amazon announced that it was replacing its forklift's batteries in 11 warehouses with hydrogen fuel cells. A Reuters news report read that the e-retail giant had acquired the right to buy up to 23% of hydrogen fuel cell maker Plug Power. Plug Power went on to say that Amazon was going to spend $70mn this year on making the switch to hydrogen power—but the end result would enable Amazon's forklifts to work round the clock, without the need to recharge. Operators will simply have to refill them with hydrogen fuel when they run out. Amazon was soon followed by Wal-Mart, and currently the entire fleet of material handling equipment at the new Wal-Mart Perishable Distribution Centre (PDC) in Balzac Alberta is powered by hydrogen fuel cells.

Toyota Motor Corp., the world's largest forklift maker, is developing hydrogen-powered models and began using prototypes at one of its plants in Japan this year. Hyster-Yale Materials Handling, another large forklift maker, bought Nuvera Fuel Cells in 2014 and has begun incorporating the technology into its products. Hyster is also looking into bio fuels and CNG options to power forklifts.

Hydrogen Fuel Cell Forklist

Hydrogen fuel cell Forklist last longer, have more power for a longer period of time and to top it all off, they refuel in less than 3 minutes, just like a car.

Here are some real world examples of why companies have been making the switch:

- Nestle Waters

  Nestle Waters compared hydrogen fuel cells to other technologies to find that, once again, hydrogen fuel cells have lower operating costs and increase worker productivity.

- Hydrogenics at GM and FedEx

  With no maintenance downtime and finishing at the same high level of performance, Hydrogenics' forklifts complete hugely successful pilot programs at GM and FedEx.

- Walmart

  Walmart will save $2 million over seven years and reduce green house gas emission by 530 tons each year by utilizing hydrogen fuel cell technology.

- Wegmans

  "It enabled us to do something that will improve job satisfaction for our employees, benefit the environment, and is right for our business." - Mike Cullen, Wegmans' Vice President of Distribution Planning.

- Whole Foods

  Whole Foods saves about 3750 man hours per year per forklift because they now do not have to change batteries at the end of every work shift.

- Central Grocers

  Central Grocers of the Chicago area purchased 140 hydrogen fuel cell forklifts which now handle all of their goods. If that weren't enough, they are purchasing 80 more this year and, of course, cutting carbon emissions along the way.

- Coca-Cola

  The second largest Coca-Cola bottler in the US is using hydrogen fuel cell-powered forklifts.

- Raymond Corporation

  Take a look at Raymond Corporation's analysis of hydrogen fuel cell-powered forklifts to conventional battery-powered forklifts. They ran the numbers, weighed the pros and cons, and chose hydrogen fuel cells.

The many Advantages and Few Disadvantages of Hydrogen Power Forklifts

Advantages

Increased Productivity – Less Downtime

Current battery electric forklifts require batteries to be changed once they run out of power; this change can take up to 30 minutes each time it is required depending on the battery size and recharging station setup. In a full days operation, depending on the size of the batteries used and how heavy the workload is there could be multiple periods of downtime due to battery changes.

Hydrogen powered forklifts can be fully fuelled in under 5 minutes in the same fashion as internal combustion forklifts, overall reducing downtime and increasing productivity.

Low Input Costs

The price of natural gas which hydrogen fuel is derived from is currently at record lows. Lower prices and increased availability means hydrogen powered forklifts are becoming a viable option for more fleets.

Smaller Equipment Footprint

Depending on the size of the forklift fleet, a large battery room may need to be operated to keep enough batteries charged to allow continuous operation. Due to the large and complex equipment needed to charge batteries on a large scale, battery rooms often occupy large amounts of floor space in warehouses. The rooms must be located in close proximity to the work floor to ensure forklifts are able to make it back to have their battery changed before it runs empty. The negative by product of having to have a large battery room powering a fleet is that less floor space is available for the storage of stock.

Hydrogen systems require a compact hydrogen generator indoors and additional equipment that can be stored outside, thus providing a smaller footprint and allowing warehouse floor space to be reclaimed for the purpose of storing additional stock.

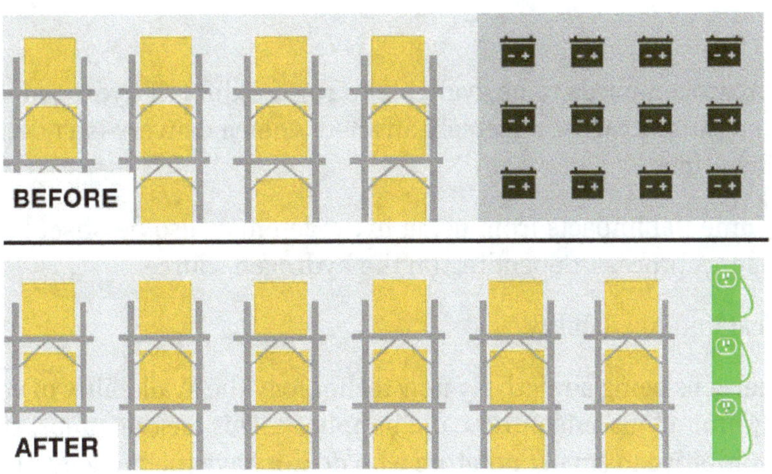

Constant Power Supply

Battery powered forklifts experience a drop in performance during the last half an hour of operation. In contrast hydrogen fuel cells experience no drops in performance across an entire shift. Hydrogen fuel cells offer sustained performance and also improve component longevity and motor wear and tear.

Lower Environmental Impacts

Hydrogen fuel cells produce zero harmful emissions, as the only by products are water vapour and heat. Companies who adopted hydrogen fuel cells can see a substantial reduction in greenhouse gas emissions. This reduction may be critical for socially responsible companies.

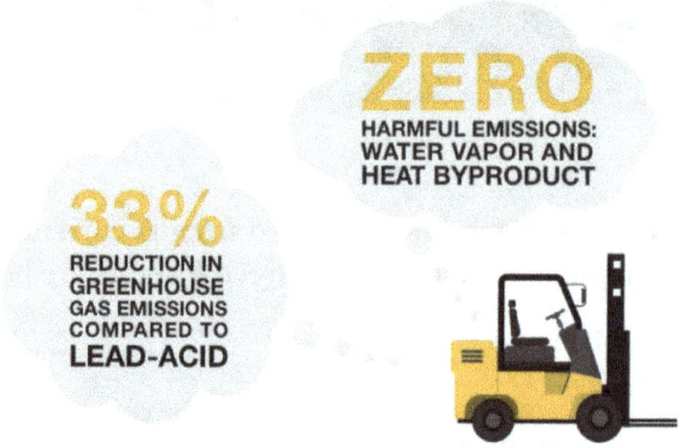

Reduced Disposal and Renewal Costs and Impacts

As opposed to batteries which must be replaced after a specific number of charging cycles, usually every 3 to 4 years, hydrogen fuels cells only need to be replaced every 10 years.

Lead batteries are complicated to dispose of due to their design and makeup, with a number of dangerous materials used in their manufacturing. Disposing of these may result in additional expenses for companies to have them safely disposed.

Disadvantages

Hydrogen Supply

A stable supply of hydrogen needs to be available to run forklifts. If hydrogen is sourced from an external supplier, operations can be adversely affected when a delivery isn't delivered resulting in extended forklift downtime.

The positive environmental impacts from using hydrogen may also be offset by the carbon emissions from the transport process, depending on the hydrogen source.

New Technology Cost and Availability

Due to hydrogen fuel cells being a relatively new technology the availability of equipment may not be widespread with limited manufacturers and suppliers. This lack of supply may also make the technology costs prohibitive to small operators who do not have the budget to cover the costs.

Fuel Cell Electric Bus

A fuel cell electric bus (FCEB) is a battery hybrid electric bus with a lightweight, on board, fuel cell power system. Through an electrochemical reaction of hydrogen and air, the fuel cells generate DC power for the bus electric drive and recharge the batteries. This provides continuous, reliable power for up to 18 hours, with no emissions except water vapor.

Greater Range, Higher Performance and Route Flexibility

FCEBs offer the range and flexibility to directly replace diesel/CNG buses:

- Range of up to 300 miles or 450 km between refuelling.
- Climbing and cold weather performance similar to diesel and CNG buses.
- Route flexibility: depot refuelling means there is no need for en-route charging infrastructure.
- Fast refuelling: 8-12 minutes of refuelling delivers 18 hours of continuous service without the need for overnight charging.
- Passenger and driver comfort with smooth, near-silent operation.

Refuelling

FCEBs are refuelled in minutes. The bus pulls into a fuelling station for an 8 to 12-minute fill-up, similar to refuelling a CNG bus. Fuelling stations are easily scalable, like CNG stations, and can be installed at existing bus depots.

The hydrogen for the fuel cells can be delivered to the bus depot in liquid form or as a compressed gas. Hydrogen can be generated:

- off-site from a variety of sources, including renewables.
- on-site, from electricity and water using electrolysis or reforming, including from renewable natural gas.

References

- Larminie, James (1 May 2003). Fuel Cell Systems Explained, Second Edition. SAE International. ISBN 0-7680-1259-7
- Wang, J.Y.; Wang, H.L. (2012). “Discrete approach for flow-field designs of parallel channel configurations in fuel cells”. International Journal of Hydrogen Energy. 37 (14): 10881–10897. doi:10.1016/j.ijhydene.2012.04.034
- Fuel-cell-technology-forklifts, analysis: cbnme.com, Retrieved 29 May 2018
- “LEMTA – Our fuel cells”. Perso.ensem.inpl-nancy.fr. Archived from the original on 21 June 2009. Retrieved 2009-09-21
- Wang, Shuangyin. “Polyelectrolyte Functionalized Carbon Nanotubes as Efficient Metal-free Electrocatalysts for Oxygen Reduction”. Journal of the American Chemical Society. 133: 5182–5185. doi:10.1021/ja1112904
- Benefits-of-hydrogen-fuel-cell-powered-forklifts: aalhysterforklifts.com.au, Retrieved 31 March 2018
- Srivastava, H. C. Nootan ISC Chemistry (12th) Edition 18, pp. 458–459, Nageen Prakashan (2014) ISBN 9789382319399
- Kakati, B. K.; Deka, D. (2007). “Effect of resin matrix precursor on the properties of graphite composite bipolar plate for PEM fuel cell”. Energy & Fuels. 21 (3): 1681–1687. doi:10.1021/ef0603582
- Fuel-cell-electric-bus: ballard.com, Retrieved 12 June 2018
- “Ballard Power Systems: Commercially Viable Fuel Cell Stack Technology Ready by 2010”. 29 March 2005. Archived from the original on 27 September 2007. Retrieved 2007-05-27
- A Kulkarni; FT Ciacchi; S Giddey; C Munnings; SPS Badwal; JA Kimpton; D Fini (2012). “Mixed ionic electronic conducting perovskite anode for direct carbon fuel cells”. International Journal of Hydrogen Energy. 37 (24): 19092–19102. doi:10.1016/j.ijhydene.2012.09.141

4

Electric Vehicle Technologies

Electric vehicles are powered using one or more electric motors. It may contain solar panels, a battery or an electric generator for converting fuel into electricity. This chapter provides a close examination of electric vehicle technologies and includes vital topics like electric vehicle battery, charge control, electric motor, traction motor, electric traction system, motor controller, etc. for a comprehensive understanding of the subject.

Electric Vehicle Battery

A battery is a device for storing chemical energy and converting that chemical energy into electricity. A battery is made up of one or more electrochemical cells, each of which consists of two half-cells or electrodes. One half-cell, called the negative electrode, has an overabundance of the tiny, negatively charged subatomic particles called electrons. The other, called the positive electrode, has a deficit of electrons. When the two halves are connected by a wire or an electrical cable, electrons will flow from the negative electrode to the positive electrode. We call this flow of electrons electricity. The energy of these moving electrons can be harnessed to do work - running a motor, for instance. As electrons pass to the positive side, the flow gradually slows down and the voltage of the electricity produced by the battery drops. Eventually, when there are as many electrons on the positive side as on the negative side, the battery is considered 'dead' and is no longer capable of producing an electric flow.

Electric Vehicle Battery Materials

Cobalt

Most electric vehicle batteries are lithium based and rely on a mix of cobalt, manganese, nickel, and graphite and other primary components. Some of these materials are harder to find than others, though none should be classified as "rare earth metals."

There are important issues surrounding battery production that must be acknowledged and addressed. For example, over 60% of the global supply for cobalt comes from the Democratic Republic of Congo (DRC), which has a poor human rights track record; international organizations have denounced for years the exploitative labor practices involved in cobalt production.

Solutions for EV Battery Sourcing Issues

First, companies must be held accountable for enacting and enforcing policies to only use ethically-sourced materials. Some companies are off to a good start. Tesla, for example, has committed to sourcing materials only from North America for its battery production facility and battery supplier LG Chem claims they have stopped using conflict-sourced cobalt.

Second, recycling can help reduce the need to search for battery materials. Cobalt is fully recyclable and roughly 15 percent of U.S. cobalt consumption is from recycled scrap today.

Third, battery technology is continuing to improve. Lithium-titanate and lithium-iron-phosphate, for example, are gaining importance in the EV market and don't need cobalt. Other battery chemistries that rely on magnesium, sodium, or lithium-sulfur are also gaining traction as they have the potential to beat lithium-ion batteries on energy density and cost.

Electric Vehicle Battery Cost

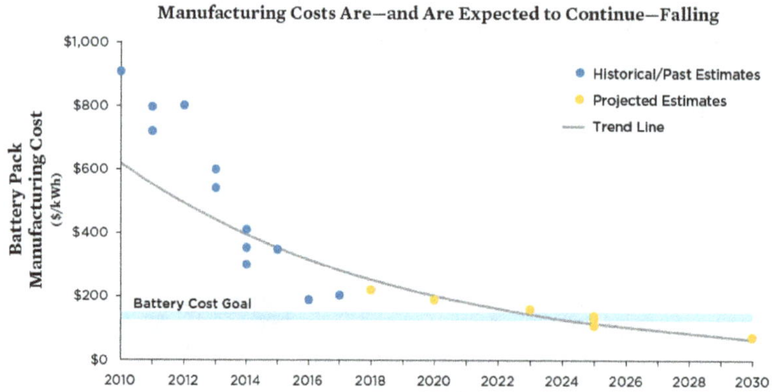

If battery costs continue to decline as EV production increases, within several years they will reach the $125–$150 target that makes EVs competitive with conventional gasoline vehicles.

If battery costs continue to decline as EV production increases, within several years they will reach the $125–$150 target that makes EVs competitive with conventional gasoline vehicles.

The price of lithium-ion batteries has fallen steeply as their production scale has increased and manufacturers have developed more cost-effective methods.

When the first mass-market EVs were introduced in 2010, their battery packs cost an estimated $1,000 per kilowatt-hour (kWh). Today, Tesla's Model 3 battery pack costs $190 per kWh, and

General Motors' 2017 Chevrolet Bolt battery pack is estimated to cost about $205 per kWh. That's a drop of more than 70% in the price per kWh in 6 years.

EVs are forecast to cost the same or less than a comparable gasoline-powered vehicle when the price of battery packs falls to between $125 and $150 per kWh. Analysts have forecast that this price parity can be achieved as soon as 2020, while other studies have forecast the price of a lithium-ion battery pack to drop to as little as $73/kWh by 2030.

Battery Types

Old: Banks of conventional lead-acid car batteries are still commonly used for EV propulsion

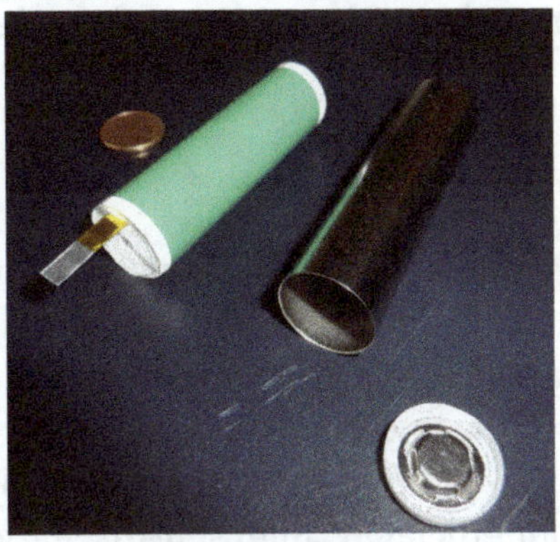

Cylindrical cell (18650) prior to assembly.

Lithium ion battery monitoring electronics (over- and discharge protection)

Lead-acid

Flooded lead-acid batteries are the cheapest and in past most common traction batteries available. There are two main types of lead-acid batteries: automobile engine starter batteries, and deep cycle batteries. Automobile alternators are designed to provide starter batteries high charge rates for fast charges, while deep cycle batteries used for electric vehicles like forklifts or golf carts, and as the auxiliary house batteries in RV's, require different multi-stage charging. No lead acid battery should be discharged below 50% of its capacity, as it shortens the battery's life. Flooded batteries require inspection of electrolyte level and occasional replacement of water which gases away during the normal charging cycle.

Traditionally, most electric vehicles have used lead-acid batteries due to their mature technology, high availability, and low cost (exception: some early EVs, such as the Detroit Electric, used a nickel–iron battery.) Like all batteries, these have an environmental impact through their construction, use, disposal or recycling. On the upside, vehicle battery recycling rates top 95% in the United States. Deep-cycle lead batteries are expensive and have a shorter life than the vehicle itself, typically needing replacement every 3 years.

Lead-acid batteries in EV applications end up being a significant (25–50%) portion of the final vehicle mass. Like all batteries, they have significantly lower energy density than petroleum fuels—in this case, 30–40 Wh/kg.While the difference isn't as extreme as it first appears due to the lighter drive-train in an EV, even the best batteries tend to lead to higher masses when applied to vehicles with a normal range. The efficiency (70–75%) and storage capacity of the current generation of common deep cycle lead acid batteries decreases with lower temperatures, and diverting power to run a heating coil reduces efficiency and range by up to 40%. Recent advances in battery efficiency, capacity, materials, safety, toxicity and durability are likely to allow these superior characteristics to be applied in car-sized EVs.

Charging and operation of batteries typically results in the emission of hydrogen, oxygen and sulfur, which are naturally occurring and normally harmless if properly vented. Early Citicar owners discovered that, if not vented properly, unpleasant sulfur smells would leak into the cabin immediately after charging.

Lead-acid batteries powered such early-modern EVs as the original versions of the EV1 and the RAV4 EV.

Nickel Metal Hydride

Nickel-metal hydride batteries are now considered a relatively mature technology. While less efficient (60–70%) in charging and discharging than even lead-acid, they have an energy density of 30–80 Wh/kg, far higher than lead-acid. When used properly, nickel-metal hydride batteries can have exceptionally long lives, as has been demonstrated in their use in hybrid cars and surviving NiMH RAV4 EVs that still operate well after 100,000 miles (160,000 km) and over a decade of service. Downsides include the poor efficiency, high self-discharge, very finicky charge cycles, and poor performance in cold weather.

GM Ovonic produced the NiMH battery used in the second generation EV-1, and Cobasys makes a nearly identical battery (ten 1.2 V 85 Ah NiMH cells in series in contrast with eleven cells for Ovonic battery). This worked very well in the EV-1. Patent encumbrance has limited the use of these batteries in recent years.

Zebra

The sodium or "zebra" battery uses a molten chloroaluminate sodium ($NaAlCl_4$) as the electrolyte. This chemistry is also occasionally referred to as "hot salt". A relatively mature technology, the Zebra battery has an energy density of 120Wh/kg and reasonable series resistance. Since the battery must be heated for use, cold weather doesn't strongly affect its operation except for in increasing heating costs. They have been used in several EVs. Zebras can last for a few thousand charge cycles and are nontoxic. The downsides to the Zebra battery include poor power density (<300 W/kg) and the requirement of having to heat the electrolyte to about 270° C (520° F), which wastes some energy and presents difficulties in long-term storage of charge.

Zebra batteries have been used in the Modec commercial vehicle since it entered production in 2006.

Lithium-ion

Lithium-ion (and similar lithium polymer) batteries, widely known via their use in laptops and consumer electronics, dominate the most recent group of EVs in development. The traditional lithium-ion chemistry involves a lithium cobalt oxide cathode and a graphite anode. This yields cells with an impressive 200+ Wh/kg energy density and good power density, and 80 to 90% charge/ discharge efficiency. The downsides of traditional lithium-ion batteries include short cycle lives (hundreds to a few thousand charge cycles) and significant degradation with age. The cathode is also somewhat toxic. Also, traditional lithium-ion batteries can pose a fire safety risk if punctured or charged improperly. These laptop cells don't accept or supply charge when cold, and so heaters can be necessary in some climates to warm them. The maturity of this technology is moderate. The Tesla Roadster (2008) uses "blades" of traditional lithium-ion "laptop battery" cells that can be replaced individually as needed.

Most other EVs are utilizing new variations on lithium-ion chemistry that sacrifice energy and power density to provide fire resistance, environmental friendliness, very rapid charges (as low as a few minutes), and very long lifespans. These variants (phosphates, titanates, spinels, etc.) have been shown to have a much longer lifetime, with A123 expecting their lithium iron phosphate batteries to last for at least 10+ years and 7000+ charge cycles, and LG Chem expecting their lithium-manganese spinel batteries to last up to 40 years.

Much work is being done on lithium ion batteries in the lab. Lithium vanadium oxide has already made its way into the Subaru prototype G4e, doubling energy density. Silicon nanowires, silicon nanoparticles, and tin nanoparticles promise several times the energy density in the anode, while composite and superlattice cathodes also promise significant density improvements.

EV Parity

In 2010, battery professor Poul Norby stated that he believed that lithium batteries will need to double their energy density and bring down the price from $500 (2010) to $100 per kWh capacity in order to make an impact on petrol cars. Citigroup indicates $230/kWh.

Toyota Prius 2012 plug-in's official page declare 21 kilometres (13 mi) of autonomy and a battery capacity of 5.2 kWh with a ratio of 4 kilometres (2.5 mi)/kWh, while the Addax (2015 model) utility vehicle already reaches 110 kilometres (68.5 mi) or a ratio of 7.5 kilometers (4.6 mi)/kWh.

Battery electric cars achieve about 5 miles (8.0 km)/kWh. The Chevrolet Volt is expected to achieve 50 MPGe when running on the auxiliary power unit (a small onboard generator) – at 33% thermodynamic efficiency that would mean 12 kWh for 50 miles (80 km), or about 240 watt-hours per mile. For prices of 1 kWh of charge with various different battery technologies, "Energy/Consumer Price" column in the "Table of rechargeable battery technologies" section in the rechargeable battery article.

United States Secretary of Energy Steven Chu predicted costs for a 40-mile range battery will drop from a price in 2008 of $12K to $3,600 in 2015 and further to $1,500 by 2020. Li-ion, Li-poly, Aluminium-air batteries and zinc-air batteries have demonstrated energy densities high enough to deliver range and recharge times comparable to conventional fossil fuelled vehicles.

Cost Parity

Different costs are important. One issue is purchase price, the other issue is total cost of ownership. As of 2015, electric cars are more expensive to initially purchase, but cheaper to run, and in at least some cases, total cost of ownership may be lower.

New PEVs would become cost efficient to consumers if battery prices would decrease from $1300/kWh to about $500/kWh (so that the battery may pay for itself).

In 2010, the Nissan Leaf battery pack was reportedly produced at a cost of $18,000. Nissan's initial production costs at the launch of the Leaf were therefore about $750 per kilowatt hour (for the 24 kWh battery).

In 2012, McKinsey Quarterly linked battery prices to gasoline prices on a basis of 5-year total cost of ownership for a car, estimating that $3.50/gallon equate to $250/kWh. In 2017 McKinsey estimated that electric cars are competitive at a battery pack cost of $100/kWh (around 2030), and expects pack costs to be $190/kWh by 2020.

In October 2015, car maker GM revealed at their annual Global Business Conference that they expected a price of $145 per kilowatt hour for Li-ion cells entering 2016.

Range Parity

Driving range parity means that the electric vehicle has the same range than an average all-combustion vehicle (500 kilometres or 310 miles), with 1+ kWh/kg batteries. Higher range means that the electric vehicles would run more kilometers without recharge.

Japanese and European Union officials are in talks to jointly develop advanced rechargeable batteries for electric cars to help nations reduce greenhouse-gas emissions. Developing a battery that can power an electric vehicle 500 kilometres (310 mi) on a single charging is feasible, said Japanese battery maker GS Yuasa Corp. Sharp Corp and GS Yuasa are among Japanese solar-power cell and battery makers that may benefit from cooperation.

- The lithium-ion battery in the AC Propulsion tzero provides 400 to 500 km (200 to 300 mi) of range per charge (single charge range). The list price of this vehicle when it was released in 2003 was $220,000.
- Driving in a Daihatsu Mira equipped with 74 kWh lithium ion batteries, the Japan EV Club has achieved a world record for an electric car: 1,003 kilometres (623 mi) without recharging.
- Zonda Bus, in Jiangsu, China offers the Zonda Bus New Energy with a 500-kilometre (310 mi) only-electric range.
- Tesla Model S with 85 kWh battery has a range of 510 km (320 miles). Tesla Model S has been built since 2012. It is priced around US$100,000.
- The supercar Rimac Concept One with 82 kWh battery has a range of 500 km. The car is built since 2013.
- The pure electric car BYD e6 with 60 kWh battery has a range of 300 km.

Specifics

Internal Components

Battery pack designs for Electric Vehicles (EVs) are complex and vary widely by manufacturer and specific application. However, they all incorporate a combination of several simple mechanical and electrical component systems which perform the basic required functions of the pack.

Battery pack on the roof of a battery electric bus

Electric truck e-Force One. Battery pack between the axles.

The actual battery cells can have different chemistry, physical shapes, and sizes as preferred by various pack manufacturers. Battery pack will always incorporate many discrete cells connected in series and parallel to achieve the total voltage and current requirements of the pack. Battery packs for all electric drive EVs can contain several hundred individual cells.

To assist in manufacturing and assembly, the large stack of cells is typically grouped into smaller stacks called modules. Several of these modules will be placed into a single pack. Within each module the cells are welded together to complete the electrical path for current flow. Modules can also incorporate cooling mechanisms, temperature monitors, and other devices. In most cases, modules also allow for monitoring the voltage produced by each battery cell in the stack by the Battery Management System (BMS).

The battery cell stack has a main fuse which limits the current of the pack under a short circuit condition. A "service plug" or "service disconnect" can be removed to split the battery stack into two electrically isolated halves. With the service plug removed, the exposed main terminals of the battery present no high potential electrical danger to service technicians.

The battery pack also contains relays, or contactors, which control the distribution of the battery pack's electrical power to the output terminals. In most cases there will be a minimum of two main relays which connect the battery cell stack to the main positive and negative output terminals of the pack, those supplying high current to the electrical drive motor. Some pack designs will include alternate current paths for pre-charging the drive system through a pre-charge resistor or for powering an auxiliary buss which will also have their own associated control relays. For obvious safety reasons these relays are all normally open.

The battery pack also contains a variety of temperature, voltage, and current sensors. Collection of data from the pack sensors and activation of the pack relays are accomplished by the pack 's Battery Monitoring Unit (BMU) or Battery Management System (BMS). The BMS is also responsible for communications with the world outside the battery pack.

Charging

Batteries in BEVs must be periodically recharged. BEVs most commonly charge from the power grid (at home or using a street or shop recharging point), which is in turn generated from a variety of domestic resources, such as coal, hydroelectricity, nuclear and others. Home or grid power, such as photovoltaic solar cell panels, microhydro or wind may also be used and are promoted because of concerns regarding global warming.

With suitable power supplies, good battery lifespan is usually achieved at rates not exceeding "0.5*C*" or so, taking two to three hours for a full charge, but faster charging can be done.

Charging time is often limited by the capacity of the grid connection. A normal household outlet delivers 1.5 kilowatts (in the US, Canada, Japan, and other countries with 110 volt supply) and 3 kilowatts (in countries with 240 V supply).

In 1995, some charging stations charged BEVs in one hour. In November 1997, Ford purchased a fast-charge system produced by AeroVironment called "PosiCharge" for testing its fleets of Ranger EVs, which charged their lead-acid batteries in between six and fifteen minutes. In February 1998, General Motors announced a version of its "Magne Charge" system which could recharge NiMH batteries in about ten minutes, providing a range of sixty to one hundred miles.

In 2005, handheld device battery designs by Toshiba were claimed to be able to accept an 80% charge in as little as 60 seconds. Scaling this specific power characteristic up to the same 7 kilowatt-hour EV pack would result in the need for a peak of 340 kilowatts of power from some source for those 60 seconds. It is not clear that such batteries will work directly in BEVs as heat build-up may make them unsafe.

Recharging Time

Electric cars like Tesla Model S, Renault Zoe, BMW i3, etc. can recharge their batteries at quick charging stations within 30 minutes to 80 percent.

In laboratories the company StoreDot from Israel reportedly first lab samples of unspecified batteries can (battery capacity in the area around 1 Ah) with as of April 2014, loaded in 30 seconds in mobile phones.

Researchers from Singapore have 2014 developed a battery that can be recharged after 2 minutes to 70 percent. The batteries rely on lithium-ion technology. However, the anode and the negative pole in the battery is no longer made of graphite, but a titanium dioxide gel. The gel accelerates the chemical reaction significantly, thus ensuring a faster charging. In particular, these batteries are to be used in electric cars. Already in 2012 researchers at the Ludwig-Maximilian-University in Munich have discovered the basic principle.

Scientists at Stanford University in California have developed a battery that can be charged within one minute. The anode is made of aluminum and the cathode made of graphite.

The electric car Volar-e of the company Applus + IDIADA, based on the Rimac Concept One, contains lithium iron phosphate batteries that can be recharged in 15 minutes.

According to the manufacturer BYD the lithium iron phosphate battery of the electric car e6 is charged at a fast charging station within 15 minutes to 80%, after 40 minutes at 100%.

Connectors

The charging power can be connected to the car in two ways. The first is a direct electrical connection known as conductive coupling. This might be as simple as a mains lead into a weatherproof socket through special high capacity cables with connectors to protect the user from high voltages. The modern standard for plug-in vehicle charging is the SAE 1772 conductive connector in the US. The ACEA has chosen the VDE-AR-E 2623-2-2 for deployment in Europe, which, without a latch, means unnecessary extra power requirements for the locking mechanism.

The second approach is known as inductive charging. A special 'paddle' is inserted into a slot on the car. The paddle is one winding of a transformer, while the other is built into the car. When the paddle is inserted it completes a magnetic circuit which provides power to the battery pack. In one inductive charging system, one winding is attached to the underside of the car, and the other stays on the floor of the garage. The advantage of the inductive approach is that there is no possibility of electrocution as there are no exposed conductors, although interlocks, special connectors and ground fault detectors can make conductive coupling nearly as safe. Inductive charging can also reduce vehicle weight, by moving more charging componentry offboard. An inductive charging proponent from Toyota contended in 1998, that overall cost differences were minimal, while a conductive charging proponent from Ford contended that conductive charging was more cost efficient.

Recharging Spots

In France, Électricité de France (EDF) and Toyota are installing recharging points for PHEVs on roads, streets and parking lots. EDF is also partnering with Elektromotive, Ltd. to install 250 new charging points over six months from October 2007 in London and elsewhere in the UK. Recharging points also can be installed for specific uses, as in taxi stands.

Travel Range before Rechargings

The range of a BEV depends on the number and type of batteries used. The weight and type of vehicle as well as terrain, weather, and the performance of the driver also have an impact, just as they do on the mileage of traditional vehicles. Electric vehicle conversion performance depends on a number of factors including the battery chemistry:

- Lead-acid batteries are the most available and inexpensive. Such conversions generally have a range of 30 to 80 km (20 to 50 mi). Production EVs with lead-acid batteries are capable of up to 130 km (80 mi) per charge.

- NiMH batteries have higher energy density than lead-acid; prototype EVs deliver up to 200 km (120 mi) of range.
- New lithium-ion battery-equipped EVs provide 320–480 km (200–300 mi) of range per charge. Lithium is also less expensive than nickel.
- Nickel-zinc battery are cheaper and lighter than Nickel-cadmium batteries. They are also cheaper (but not as light) than Lithium-Ion batteries.

Finding the economic balance of range versus performance, battery capacity versus weight, and battery type versus cost challenges every EV manufacturer.

With an AC system or Advanced DC systems regenerative braking can extend range by up to 50% under extreme traffic conditions without complete stopping. Otherwise, the range is extended by about 10 to 15% in city driving, and only negligibly in highway driving, depending upon terrain.

BEVs (including buses and trucks) can also use genset trailers and pusher trailers in order to extend their range when desired without the additional weight during normal short range use. Discharged baset trailers can be replaced by recharged ones in a route point. If rented then maintenance costs can be deferred to the agency.

Such BEVs can become Hybrid vehicles depending on the trailer and car types of energy and powertrain:

- The Tesla Roadster (build 2008–2012) can travel 245 miles (394 km) per charge.
- Tesla Model S with 85 kWh battery has a range of 510 km (320 miles). Tesla Model S has been built since 2012. It is priced around US$100,000.
- The supercar Rimac Concept One with 82 kWh battery has a range of 500 km. The car is built since 2013.
- The pure electric car BYD e6 with 60 kWh battery has a range of 300 km.
- The bestseller Nissan Leaf model year 2016 with 30 kWh battery has a range of 172 km.

Trailers

Axillary battery capacity carried in trailers can increase the overall vehicle range, but also increases the loss of power arising from aerodynamic drag, increases weight transfer effects and reduces traction capacity.

Thermal Effects

The internal resistance of some batteries may be significantly increased at low temperature which can cause noticeable reduction in the range of the vehicle and on the lifetime of the battery.

Swapping and Removing

An alternative to recharging is to exchange drained or nearly drained batteries (or battery range extender modules) with fully charged batteries. This is called battery swapping and is done in exchange stations.

On the other hand, MIRA has announced a retrofit hybrid conversion kit that provides removable battery packs that plug into a wall outlet for charging. Also XP Vehicles uses extension-cord-free charging hot-swap battery (removable power pack to recharge at home without extension cord).

Features of swap stations include:

1. The consumer is no longer concerned with battery capital cost, life cycle, technology, maintenance, or warranty issues;
2. Swapping is far faster than charging: battery swap equipment built by the firm Better Place has demonstrated automated swaps in less than 60 seconds;
3. Swap stations increase the feasibility of distributed energy storage via the electric grid.

Concerns about swap stations include:

1. Potential for fraud (battery quality can only be measured over a full discharge cycle; battery lifetime can only be measured over repeated discharge cycles; those in the swap transaction cannot know if they are getting a worn or reduced effectiveness battery; battery quality degrades slowly over time, so worn batteries will be gradually forced into the system).
2. Manufacturers unwillingness to standardize battery access/implementation details.
3. Safety concerns.

Re-filling

Zinc-bromine flow batteries can be re-filled using a liquid, instead of recharged by connectors, saving time.

Leasing

Three companies are working on battery lease plans. Greenstop has completed trials of their ENVI Grid Network which allows consumers to easily monitor and recharge electric vehicle batteries. Think Car USA plans to lease the batteries for its City electric car to go on sale next year. Better Place is creating a system for consumers to "subscribe" to a service that offers recharging stations and battery exchange.

Electric utilities are considering plans which would include providing electric vehicles to users (at a low price) and get their profits from selling the energy.

V2G and Afteruse

Smart grid allows BEVs to provide power to the grid at any time, especially:

- During peak load periods (When the selling price of electricity can be very high. These vehicles can then be recharged during off-peak hours at cheaper rates while helping to absorb excess night time generation. Here the vehicles serve as a distributed battery storage system to buffer power).

- During blackouts, as backup.

Pacific Gas and Electric Company (PG&E) has suggested that utilities could purchase used batteries for backup and load levelling purposes. They state that while these used batteries may be no longer usable in vehicles, their residual capacity still has significant value.

Lifespan

Individual batteries are usually arranged into large battery packs of various voltage and ampere-hour capacity products to give the required energy capacity. Battery service life should be considered when calculating the extended cost of ownership, as all batteries eventually wear out and must be replaced. The rate at which they expire depends on a number of factors.

The depth of discharge (DOD) is the recommended proportion of the total available energy storage for which that battery will achieve its rated cycles. Deep cycle lead-acid batteries generally should not be discharged to below 20% of total capacity. More modern formulations can survive deeper cycles.

In real world use, some fleet Toyota RAV4 EVs, using Nickel–metal hydride battery, have exceeded 100,000 miles (160,000 km) with little degradation in their daily range. Quoting that report's concluding assessment:

- "The five-vehicle test is demonstrating the long-term durability of Nickel Metal Hydride batteries and electric drive trains. Only slight performance degradation has been observed to-date on four out of five vehicles. EVTC test data provide strong evidence that all five vehicles will exceed the 100,000-mile (160,000 km) mark. SCE's positive experience points to the very strong likelihood of a 130,000 to 150,000-mile (240,000 km) Nickel Metal Hydride battery and drive-train operational life. EVs can therefore match or exceed the lifecycle miles of comparable internal combustion engine vehicles.
- "In June 2003 the 320 RAV4 EVs of the SCE fleet were used primarily by meter readers, service managers, field representatives, service planners and mail handlers, and for security patrols and carpools. In five years of operation, the RAV4 EV fleet had logged more than 6.9 million miles, eliminating about 830 tons of air pollutants, and preventing more than 3,700 tons of tailpipe carbon dioxide emissions. Given the successful operation of its EVs to-date, SCE plans to continue using them well after they all log 100,000-miles."

Lithium ion batteries are perishable to some degree; they lose some of their maximum storage capacity per year even if they are not used. Nickel metal hydride batteries lose much less capacity and are cheaper for the storage capacity they give, but have a lower total capacity initially for the same weight.

Jay Leno's 1909 Baker Electric still operates on its original Edison cells. Battery replacement costs of BEVs may be partially or fully offset by the lack of regular maintenance such as oil and filter changes required for ICEVs, and by the greater reliability of BEVs due to their fewer moving parts. They also do away with many other parts that normally require servicing and maintenance in a regular car, such as on the gearbox, cooling system, and engine tuning. And by the time batteries do finally need definitive replacement, they can be replaced with later generation ones which may offer better performance characteristics.

Lithium iron phosphate batteries reach according to the manufacturer more than 5000 cycles at respective depth of discharge of 70%. As the world's largest manufacturer of lithium iron phosphate batteries are concerned BYD, which has developed through precision manufacturing a wide range of cells for deep cycle applications, such as in use in stationary storage systems. After 7500 cycles with discharge of 85% this still have a spare capacity of at least 80% at a rate of 1 C; which corresponds with a full cycle per day to a lifetime of min. 20.5 years. The lithium iron phosphate battery Sony Fortelion has after 10,000 cycles at 100% discharge level still a residual capacity of 71%. This accumulator is since 2009 on the market.

Used in solar batteries Lithium-ion batteries have partly a very high cycle resistance of more than 10,000 charge and discharge cycles and a long service life of up to 20 years.

Plug-in America has among drivers of the Tesla Roadster (2008), a survey carried out with respect to the service life of the installed battery. It was found that after 100,000 miles = 160,000 km, the battery still had a remaining capacity of 80 to 85 percent. This was regardless of in which climate zone the car is moved. The Tesla Roadster was built and sold between 2008 and 2012. For its 85-kWh batteries in the Tesla Model S Tesla are 8-year warranty with unlimited mileage.

Varta Storage abandons his family engion family and engion home a guarantee of 14,000 full cycles and a service life of 10 years.

As of December 2016, the world's all-time best-selling electric car is the Nissan Leaf, with more than 250,000 units sold since its inception in 2010. Nissan stated in 2015 that until then only 0.01 percent of batteries had to be replaced because of failures or problems and then only because of externally inflicted damage. There are a few vehicles that have already covered more than 200,000 km; none of these had any problems with the battery.

Recycling

At the end of their useful life, batteries can be recycled.

Safety

The safety issues of battery electric vehicles are largely dealt with by the international standard ISO 6469. This document is divided in three parts dealing with specific issues:

- On-board electrical energy storage, i.e. the battery.
- Functional safety means and protection against failures.
- Protection of persons against electrical hazards.

Firefighters and rescue personnel receive special training to deal with the higher voltages and chemicals encountered in electric and hybrid electric vehicle accidents. While BEV accidents may present unusual problems, such as fires and fumes resulting from rapid battery discharge, many experts agree that BEV batteries are safe in commercially available vehicles and in rear-end collisions, are safer than gasoline-propelled cars with a rear gasoline tanks.

Usually, battery performance testing includes the determination of:

- State Of Charge (SOC)
- State of Health (SOH)
- Energy Efficiency

Performance testing simulates the drive cycles for the drive trains of Battery Electric Vehicles (BEV), Hybrid Electric Vehicles (HEV) and Plug in Hybrid Electric Vehicles (PHEV) as per the required specifications of car manufacturers (OEMs). During these drive cycles, controlled cooling of the battery can be performed, simulating the thermal conditions in the car.

In addition, climatic chambers assure constant environmental conditions during the characterization and allow the simulation to be performed for the full automotive temperature range covering climatic conditions.

Patents

Patents may be used to suppress development or deployment of this technology. For instance, patents relevant to the use of Nickel metal hydride cells in cars were held by an offshoot of Chevron Corporation, a petroleum company, who maintained veto power over any sale or licensing of NiMH technology.

Tips to Extend Battery Life of your Electric Vehicle

Avoid full Charging when you can

One reason that batteries in mobile devices only last a couple years is that they are being pushed to their maximum capacity—frequently getting fully charged and fully drained. Consumer products are advertised by their battery operation time, not their battery lifespan. This means that every possible electron will be shoved in there. Charging to maximum capacity might give you the most possible use for that one charge, but it is one of the worst things that you can do to lithium batteries.

In the 2011 Nissan LEAF, there is a Long Battery Life setting that tells the car to stop charging at 80 percent. This reduces the available range, but could greatly increase the lifespan of your battery pack. If your normal daily driving can be done with less than an 80 percent charge or you can charge mid-day, this simple setting is one of the easiest things that you can do to increase the battery's lifespan.

One additional advantage of not charging up all the way is that it leaves room to store energy from regenerative braking. Often when the batteries are full or near full, regent will be disabled to avoid overcharging the batteries.

For Pure EVs, Avoid Deep Discharging your Battery Pack

Lithium-ion packs prefer a partial cycle rather than a deep discharge. Since lithium-ion chemistries do not have a memory effect, there is no harm using a partial discharge. Not only will this avoid excessive wear, it will also mean that—with a little planning—you will arrive at your destination with range to spare.

The Nissan LEAF has 12 "fuel bars" that tell you the charge level of the batteries. Just like you were taught to keep your hands on the wheel at 10 and 2, it is a good idea to keep your LEAF's charge level within 10 and 2 bars of charge.

EV drivers are often asked, "How far can you go on a charge?" But if you can help it, avoid trying to personally find out for yourself by driving all the way down to an empty battery to test your electric car's full range.

For Plug-in Hybrids, Consider "mountain-mode" or Leaving EV-mode at Key Times

In a plug-in hybrid like the Prius Plug-in Hybrid, or a range-extended electric vehicle such as the Chevy Volt, it's more common to use all of the available battery capacity before switching to gasoline power. Fortunately, the vehicle's battery management system knows when to stop drawing from the battery—to avoid deep cycling.

But if you know that your planned route includes a long hill climb, switch to Mountain Mode—which kicks on the gas engine and saves juice in the battery—at least 20 minutes before you start the climb. This will ensure that the batteries are not deep discharged during a long steep climb and you'll have all the power you need to pass as you go up the mountain.

The Toyota Prius Plug-in Hybrid gives drivers the similar option of voluntarily pushing the car out of EV mode—and for a period of time use more gas and less battery.

Use Timers to Minimize the Time Spent at a High State of Charge

You may have heard that, in many cases, it's better for the grid, the environment, and your wallet if you charge your EV at night. But careful timing of your charging schedule is also better for your batteries. Again, the idea is that lithium-ion batteries are most stable when they are about half charged—not too full and not too empty. However, a 50 percent charge may not provide adequate range for your daily driving—or may push your state of charge to the limit during your very next commute.

So how do you keep the batteries in their most stable state as long as possible while still ensuring that you have enough range when you leave in the morning? Many electric cars and charging equipment providers offer the ability to program the time of day the charging takes place. The best practice is to set the timer so the vehicle is at the charge level you need about 60 minutes before you plan to leave. Delaying the charge minimizes the time the batteries spend when the cell voltage is high.

The 60-minute buffer has two purposes. First, if something unexpected comes up, like an unscheduled meeting at work, you are not waiting for your car to charge in order to leave. Second, this time allows the batteries time to cool down after charging before they are used for driving. (This second reason is not as important during cold weather, since this warmth would help fight the cold.)

One exception to delaying the charge is if the batteries are almost completely drained. In that case, it's best to begin charging as soon as possible.

Most electric vehicles, and many of the charging equipment systems, provide the ability to schedule charging at any time.

On a Hot Day, Try to Park in the Shade. During the Winter, Park in a Garage, Rather than on the Street

Many electric cars have thermal management—via air or liquid—that is great for keeping your batteries comfortable, but it comes at a price. Running fans and AC compressors use energy. On a hot day, if you were to park your car on a hot asphalt parking lot, you would not want the thermal management system to run continually. If it did, hours later when you came back, your batteries could be somewhat drained from cooling themselves. Alternatively, the car's thermal management system could run for a few minutes after parking and then shutdown until the vehicle is restarted. That means that after that cool-down cycle completes, your batteries are slowly being heated by their surrounding. Neither of these solutions are ideal for battery longevity.

If your EV has Thermal Management and the Weather is Extreme, Plug in whenever you can

This will, of course, charge your batteries, but more importantly, it will engage the thermal management system continuously without draining the batteries. Remember that lithium batteries like the same temperature ranges that you and I do. If you had to sit in the sun in a hot parking lot, would you want the AC turned on? Yes, you would and so would your vehicle's batteries.

Plan Ahead for Period of Extended Storage

The Tesla Roadster has a "storage mode." This makes it easy: plug in the car, set it to storage mode and the vehicle's firmware takes care of the batteries. If your plug-in vehicle has a special storage mode, use it. If not, and you will not be driving your plug-in car for a month or more, here is what I recommend:

(1) Store the vehicle in a cool, but not freezing, location;

(2) Charge to 60 percent;

(3) The batteries slowly self discharge;

(4) When the charge is below 20 percent or 3 months later, whichever comes first, charge the vehicle back up to 60 percent.

With Internet-enabled EVs such as the LEAF and Volt, you can run a smartphone app and check the charge level. When the charge is below 20 percent, you can even initiate charging with these apps.

If you have your car in storage mode for a month or more, this can cause the batteries to become unbalanced. If the pack is unbalanced, your range will be greatly reduced. To alleviate this problem, simply fully charge your batteries before the next time you drive the car. Even if you generally use an 80 percent charge, the next two times you charge the car after a storage event, charge it all the way up. This will ensure the batteries are rebalanced.

To Maximize Battery Life, Minimize use of DC Quick Charge

I have saved the best for last. DC quick chargers can dump about 80 miles of range into the Nissan LEAF's battery pack in 30 minutes. Fast charging is very useful if you need to drive more than the vehicle's single-charge range in a single day. However, the batteries pay a price for this convenience.

Regular use of fast charging will cost you about 1 percent of capacity per year. For example, if you avoid fast charging, you may have 80 percent capacity after 10 years of normal use. However, if fast charging is your primary fueling method, then your capacity would be 70 percent after 10 years. While the cost to the battery's lifespan may not be as much as you may think, fast charging still takes a toll that should be avoided when possible.

Charge Control

A charge controller is an essential part of nearly all power systems that charge batteries.Electric vehicle (EV) charging can negatively impact electric distribution networks by exceeding equipment thermal ratings and causing voltages to drop below standard ranges.

Several strategies have been introduced in order to control EV charging to prevent negative impacts on the grid. These can be classified into centralized, decentralized, and autonomous charging control strategies.

Centralized Charging Control

Many methods of EV charge management have focused on centralized control methods. In centralized scheduling and control, the control algorithm of EVs is done centrally after collecting all the information about the EVs status and owners` preferences as well as other system data, such as market prices and system constraints and loading. The central controller can be an aggregator or the system operator. Centralized controllers usually result in the optimal utilization of the system resources but need a mature communication infrastructure. Figure shows a schematic for the centralized control method, where all data from the different participants in the system are sent to the main controller that processes all the data and sends back reference signals to be followed by different entities.

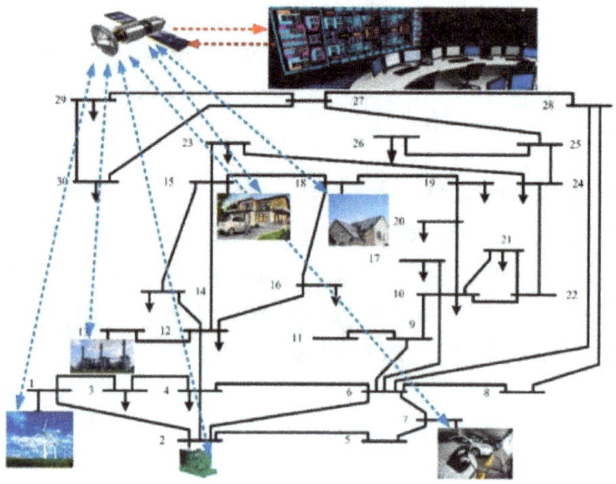

Figure: Centralized Controller schematic.

In, the authors used a control algorithm that uses the EV as another voltage control device in the system to mitigate the intermittency of distributed solar generations. The control algorithm consists of two stages where coordination between the EVs and the on-line tap changing transformer is done in the first stage. Then, in the second stage, a correction is made to the EV charging/discharging to provide a fast response to the fluctuation in the solar energy. The algorithm performed well in mitigating the overvoltage/under voltage problems in the system. However, the mobility of the EVs was not taken into consideration.

Electric vehicles on large scale were used to provide frequency regulation (FR) in the utility grid, in which the authors presented a coordinated control strategy for large-scale EVs, BESSs (Battery Energy Storage Station) and traditional FR resources involved in Automatic Generation Control (AGC). It was shown that EVs and BESS can provide fast response in case of disturbance of short periods but if the disturbance continued for longer period, only the AGC will participate in the response continuously. In, the effect of EVs in providing ancillary services with wind integration was investigated. It was shown that the regulation power requirements from conventional generators were greatly reduced with the integration of a V2G system participating in load frequency control. In, a V2G algorithm was developed to optimize energy and ancillary services scheduling for a third party, called the aggregator. An optimal bidding formulation for EVs performing regulation up and down with only unidirectional power flow was developed. The simulations were performed on a simulated market with constant prices of regulation services over the study year.

None of the V2G studies in, however, addressed charging impacts on the distribution system and they require significant communications bandwidth to frequently dispatch the EVs. Additionally, the optimization requires significant computational power by the centralized controller.

In, multi-objective optimization was used to coordinate the charging of the plug-in EVs in a way that satisfies the network technical constraints as well as customer convenience. Heavy communication between the EV customers, vehicle coordinator and Energy hub operator at the distribution system company is required. It also requires the knowledge of load profiles.

In economic-based charge control was investigated. It was found that price-based methods can sometimes cause distribution system overloads in the night hours due to low system prices but loss optimization would always flatten the load profile effectively. The method requires good communication infrastructure to perform well. In a three-level hierarchical control algorithm was proposed for coordinating EVs charging at the provincial level in China. The user preferences are sent from the lower level from the different charging stations to a higher level on the municipals level where these data are added to other non-EV loads and pricing times and finally all the data from the different municipals are sent and processed by the provincial level operator. The algorithm is good for a vertically integrated utility system, where the conventional power system architecture is still being followed. Rebound effects, where there is a sudden load increase/decrease due to the instantaneous connection/disconnection of considerable number of EVs in the low/high price period, were considered and the rebound effects were solved using the coordination from upper municipal level coordinator. However, there was no consideration for the generation and network constraints.

Three approaches were studied in; dumb charging, dual tariff policy and smart charging. Voltage profiles and lines congestion levels were evaluated, for the peak load hour, and for grid technical limits checking. Also, network lsses were evaluated for a typical daily load profile. Smart charging with hierarchical centralized control showed the best performance, and it showed that voltage is the limiting factor for higher integration of EVs.

Communication with the transformer substation for the sake of fair charging of electric vehicles as soon as possible was investigated in. The connection rate of EVs keeps increasing until the set point limit is reached and then it varies up and down to maintain the set point level. At the end of each minute interval, the EV chargers will again attempt to connect with a connection rate probability. The random process used for connecting the electric vehicles ensures that each of the EVs has fair access to the available power. In, the authors proposed a centralized charging control that allows the EVs to find, via a distributed communication network, either the closest charging station, and then only be allowed to charge if there are no network constraints, or the charging station that will allow for the quickest charge.

In, an algorithm that provides real time energy management for a grid connected charging park was proposed. The algorithm uses a fuzzy controller to yield the charging rate of each PHEV based on the current and estimated power needed by the PHEVs, the estimated power generated by the PV, and the daily energy tariff. In case of high system loading, the output of the fuzzy controller is modified to avoid negative impacts on the grid.

A market mechanism for the optimal allocation of the charging capacity to the vehicles was proposed in. The main purpose of the developed mechanism was to allow the owners of the vehicles to express their individual preference as well as ensure network stability. The author tried to open the door for multi-tiered user plans in which different users can be offered different resource allocation based on their preference and their willingness to pay. At the same time, network constraints such as total network loading, voltage drop and phase unbalance were taken into consideration.

In, a central electric vehicle charging optimization algorithm was proposed. The optimization was based on a receding horizon linear optimization problem. The authors took into consideration constraints such as the transformer and line limitations, phase unbalance and voltage stability within the network. As a simplification for the problem, the authors used a DC-equivalent model for the distribution network. Later, the algorithm was applied to the actual distribution network. In making this decision, the authors were looking not just at the current point in time, but at the best possible solution for a finite future charging horizon in discrete intervals since underlying conditions may change unexpectedly (such as vehicles arriving or departing).

Although Centralized control strategies can result in the optimal utilization of the system resources, it has a number of disadvantages:

1. The need for large investment in the communication infrastructure, especially at the distribution level.
2. An enormous number of messages needs to be communicated within a very limited period of time, which might cause communication issues, such as high latency and low quality of service.
3. High computation burden for processing a large amount of data.
4. Loss of the main communication link or problems with the central controller might have severe consequences on the system integrity.
5. User privacy issues since the central controller has access to the data of all users.

Decentralized Charging Control

Other EV charge strategies have focused on decentralized control algorithms that use reduced communications infrastructure and computational burden. In decentralized, or distributed, control, the charging/discharging reference decision is not processed and generated using the aggregator or the system operator. Instead, the operator generates some sort of a signal to incentivize the participating EVs to do certain action. However, each EV decides on its charging/discharging rate based on its own preference and no private information is sent back to aggregator or the operator. Other works use multi-agent control algorithms. A multi-agent system consists of two or more intelligent virtual or physical entities that cooperate and interact with each other to achieve a certain objectives related to their environments.

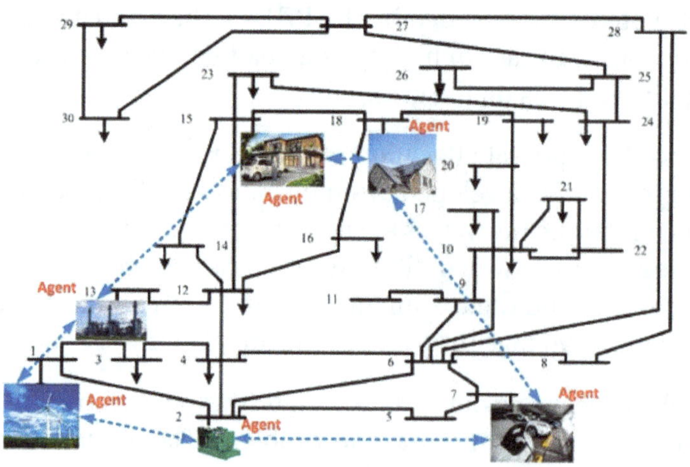

Figure: Decentralized Controller schematic.

A decentralized controller using price-driven coordination between the utility and the EVs was considered in. The authors used two gradient optimization methods, one that is based on cost and the other that is based on primal-dual approach, to minimize the total load variance in the distribution network. They assumed a uniform fleet with each car having a maximum capacity of 1.96 kW. The proposed approach changes the feedback control signal of each EV taking into consideration the state of the feeder supplying the EV. Although the method converges to a near-optimal load variance while ensuring that the feeders are not overloaded, special care should be given to the algorithm step size to ensure convergence. An electric vehicle for charging and discharging in households was considered in. In this study, the impact of price-based demand response strategies on smart household load pattern variations is assessed. A forecasting procedure using hybrid wavelet transform (WT)—ANN was considered to accurately consider the price elasticity of demand. The household load datasets are acquired to perform optimal appliance scheduling considering an hourly varying price tariff scheme. The algorithm presented in was based on EVs setting their own charge profiles according to price forecasts. In, a two-stage control algorithm was used to coordinate the charging of EVs in the presence of multiple aggregators. The algorithm takes into consideration the network constraints. A third-party price coordinator was introduced to generate a fair price signal between the distribution system operator, who tries to efficiently operate the system without violations and the aggregators, who try to charge the EVs according to their preferences and with the lowest cost. Although the method achieves satisfactory results, the used linear programming model and assumptions did not accurately characterize the charging process of the EVs by neglecting mobility aspects and battery efficiency.

In, a large population of plug-in electric vehicles was used to mitigate wind intermittency and frequency regulation. Each EV adjusts its charging or discharging power in response to a communal virtual price signal and based on its own urgency level of charging. The proposed scheme created a cost saving opportunity for both the EV owner and the utility. In, a pricing scheme that conveys price and quantity information to the load aggregator (LA) was developed and compared with the pricing only scheme. It was shown that the price/quantity scheme is insensitive to the regularization penalty and requires less computation capability than the pricing only scheme. The pricing scheme was to minimize the charging costs for the EV owner. If the objective was changed to profit maximizing for the utility, the method loses its beneficial properties.

Game theory has been recently used by multiple researchers to coordinate the charging process of EVs by achieving the Nash equilibrium where no player has anything to gain by changing his/her own strategy. In, the coordination of EVs was performed using non-cooperative games to minimize generation cost. In optimal demand-side management (DSM) is achieved using a model derived from game-theory. Each consumer's scheduler is required to broadcast its consumption schedule to all other participants in the DSM program. In, a mean field game theoretic approach was proposed to control the charging of the EVs without increasing the grid peak and taking the customers` preferences into consideration.

In, a distributed framework was suggested to charge the EVs at comparable charging rates without overloading the upstream service transformer. In, two electric vehicle charging algorithms were proposed, one centralized and one distributed, and their performances in simulations that used real vehicle data were compared, on a model that is based on a real low voltage network in northern Melbourne, Australia. The proposed algorithm for distributed charging used probability criteria to decide whether the vehicle will be charged or not. This probability is based on the node voltage and the state of charge (SOC) of the EV's battery. It does not take into consideration the maximum charging rate. The method can be used only up to 25% penetration level and it is sensitive to the location of vehicles in the network; when vehicles are connected near the far ends of the network, there is a significantly increased risk of voltage drop.

In, control of energy flow between EVs and the grid has been demonstrated using fuzzy logic controllers (FLC) mainly for voltage compensation and load flattening. The proposed technique assumes that the EVs are available to be charged and discharged. It also assumes that all EVs in a certain area will charge from a certain charging station and the charging station will work most likely as an aggregator. In, the effectiveness of distributed additive increase and multiplicative decrease (AIMD) charging algorithms at mitigating the impact of domestic charging of EVs on low-voltage distribution networks was investigated. The proposed method tries to achieve fair charging among different vehicles without violating the voltage constraint or overloading the substation and at the same time it took into consideration, the time-of-use prices. To achieve the mentioned objectives, a simple radial communication between the distribution station and the vehicles is used.

These, and other similar decentralized charging methods, rely on reduced communications infrastructure from the utility or the aggregator, even if there is no communication from the EVs back to the grid. Although, the decentralized control algorithms reduce the need for expensive communication infrastructure and the computation burden, they have the following disadvantages:

- They do not always ensure optimality and best use of resources.
- They may result in a rebound effect, which can be harmful to the system.
- They have a limited ability in participating in ancillary service markets.
- They are vulnerable to changes in customers' behavior.

Autonomous Charging Control

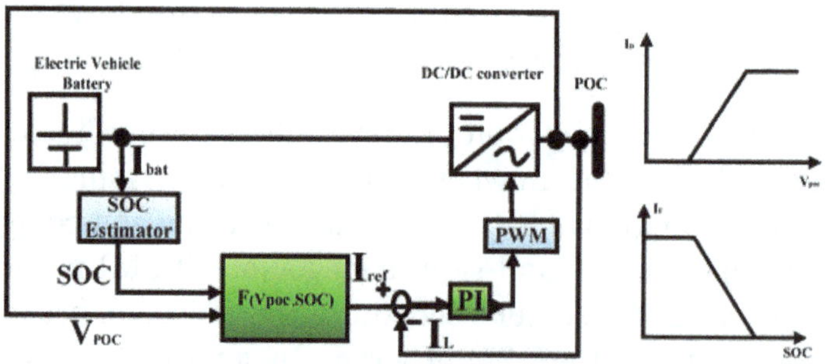

Figure: Autonomous Droop Controller (POC: point of charging).

Autonomous charging control algorithms were considered in the literature as a part of decentralized algorithms but with no communication at all. However, due to the increased work in that type of controllers, we preferred to consider it as a third type of control algorithms. Autonomous controllers can be the first step for integrating a considerable number of EVs in the absence of communication infrastructure in the distribution system in many countries. An autonomous controller mainly depends on local inputs to decide on the charging/discharging rate. Also, an autonomous controller can be viewed as a lower layer of a more comprehensive decentralized strategy, where decisions should be taken on the secondary distribution system instantaneously. An example of an autonomous droop controller is shown in figure.

In voltage-constrained local optimization of EV charging was suggested. Each EV in the system optimizes its own charging aiming to maximize its charging rate while not violating nodal voltage or feeder loading constraints. However, it does not consider charging fairness and battery SOC of each EV. In fact, comparable performance can be obtained using simpler control structures.

In an autonomous controller, provision of ancillary services such as secondary and tertiary frequency regulation are harder to do due to the absence of communication, hence the direct coordination of the charging/discharging of EVs. However, the primary frequency and voltage regulations are possible using the droop controllers. A voltage-feedback control structure for EVs in a distribution system was suggested in. However, the issue of fairness among EVs connected to different buses in the system was not addressed in. SOC dependency of charging rate was not considered, either. Fair charging means that EVs with similar initial SOC should charge at the same charging rate irrespective of their location in the distribution system. Since some of the EVs are connected to the upstream buses near the substation, they have the advantage of having a higher charging rate due to the higher voltage than those connected to downstream buses. In, the authors propose an active power/frequency (P–f) droop control strategy to be implemented at the EV coupling inverter where the EV will autonomously adapt its power output based on the microgrid (MG) frequency. This work proposed the use of this control strategy during service restoration. In, The authors proposed an autonomous distributed V2G control scheme where electric vehicles supply a distributed spinning reserve according to the frequency deviation at the plug-in terminal. The authors in proposed a method that uses only local information which is the node voltage, SOC and time of required charging given by the user and based on that information, the algorithm controls the charging rate. It uses an averaging technique method to find the set point voltages used in the controller that mainly depends on historical data. Fairness was not completely proved in the results. In, the author proposed an effective, autonomous, voltage-based control scheme for charging electric vehicles. This control scheme coordinates charging among the EVs connected to the distribution nodes in a fair manner so that voltage violations are avoided. The proposed method uses constant gain values and the upstream point charges faster than the downstream one. In, The author developed an autonomous voltage feedback control structure for EV charging based on the model in. This control structure relies on the local voltage measurement at the point where the EV is plugged in. It compares the system measured voltage at the point of charging with a predefined reference voltage. The drawback of the proposed algorithm is that it needs to update the reference set points with each seasonal variation, and the upstream point charges relatively faster than the downstream one. In, a new voltage based EV chargers was proposed. The controller used local voltage as one of the inputs and the state of charge as the second input. Then, a nonlinear exponential controller decides on the charging rate based on the inputs. The aim for this work was to eliminate the need for seasonal changing of the controller set-points mentioned in the previous autonomous work. This is done by keeping the voltage set-point as a constant value for all the EVs at the different buses and relaxing the controller sensitivity to the voltage to ensure fair charging among the EVs. A fuzzy-based autonomous controller was introduced in to eliminate the drawback of slow charging rate noted in and at the same time ensure a fair charging rate among the different EVs.

Autonomous charging algorithms eliminate the need for the communication infrastructure and can also be used as the first layer of hierarchical control algorithms where decisions should be done at the local levels. Also, autonomous control algorithm is the only option for utilities with no available communication infrastructure at the distribution level. However, if autonomous control algorithm is used alone, it will have the following disadvantages:

- Lack of the optimal operation of the system.
- Lack of ability to participate in ancillary service markets.
- May result in a rebound effect.
- Vulnerable to changes in customers' behavior.

Based on the above discussion about the different types of controllers that are used to manage the charging of EVs, a summary of the different papers discussed in table below. Also, proper classification of control algorithms and their compatibility with both vertically integrated and restructured utility systems can be reached.

Table: Summary of the discussed control techniques.

Model	Type	Capacity (kWh)	Charging Rate (kW)	Electric Range (miles)	Price ($)
Nissan Leaf	BEV	30	6.6	107	$29,000
Tesla model S	BEV	100	10	315	$71,000
Chevrolet Bolt	BEV	60	7.2	238	$37,500
Toyota Prius	PHEV	9	3.3	25	$28,000
Ford Fusion Energy	PHEV	7	3.3	19	$33,900

Depending on the utility structure, communication infrastructure, electricity market design and the level of sector liberalization, one or more of the controllers` types can be used. For example, multiple central aggregators can be used in a restructured utility system to achieve the reliable and cost-effective operation of their zones or the group of entities they have access to their information. Also, one central controller can be used to control the vertically integrated system.

As the future electric utilities move toward more deregulated ones, decentralized and multi-agent systems will be more suitable due to their several advantages. For instance, current demand side management programs can send minimum signals to provide guidance to a decentralized controller that makes the final decision based on the local level.

Autonomous controllers, on the other hand, can be the first step for integrating considerable number of EVs in the absence of communication infrastructure in the distribution system in many countries. Also, having autonomous control at the lowest level of the centralized/decentralized controller is advantageous since it reduces the communications traffic to the V2G aggregator.

Directive Warning Sounds

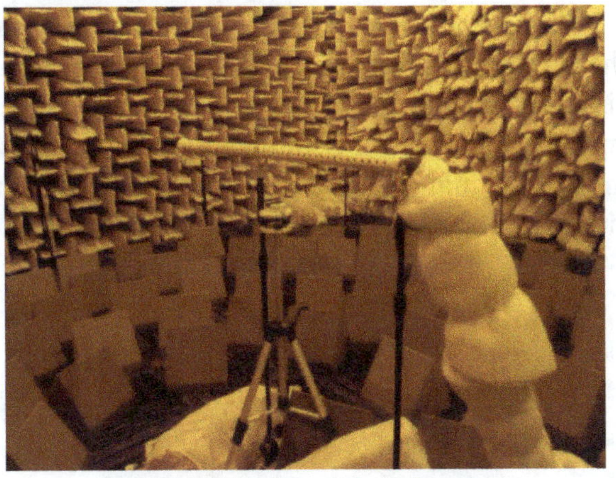

Measuring in the anechoic chamber

Increasing demand from both consumers and legislators for more fuel-efficient vehicles has increased the production of both hybrid and electric vehicles. When driving in electric mode at speeds below around 20 mph, where the level of road-tyre noise is low, these vehicles are significantly quieter than the alternative internal combustion engine vehicles. This is a significant benefit for environmental noise pollution in urban areas; however, the noise emitted by internal combustion engine vehicles is also an important warning signal for pedestrians, particularly those who are visually impaired, and cyclists to indicate that a vehicle is approaching. Therefore, there is currently concern that the lack of noise emissions from electric vehicles may cause a safety risk for pedestrians and cyclists and consequently legislation is being introduced to ensure that hybrid and electric vehicles are audible.

Vehicle-mounted measurement

To ensure the safe operation of hybrid and electric vehicles at low speeds it may be necessary to generate an external warning sound using a loudspeaker system. By designing the directivity of the sound field reproduced by the warning system, it may be possible to reach a compromise between providing sufficient audible queues to pedestrians and cyclists and avoiding unnecessary noise pollution. This project, which was part of an MEng Acoustical Engineering Group Design Project, investigated methods of implementing a directional warning sound system for electric and hybrid vehicles.

A directional sound field can be readily produced using an array of loudspeakers; however, for the automotive application the cost and weight of such a system are too high. Therefore, a low-cost directional sound radiator was investigated which uses a single loudspeaker unit coupled with a passive radiating unit which consists of a pipe in which a line of holes is drilled. The individual holes act as sound sources and the system thus produces a directional response similarly to a loudspeaker array. The directivity of this system can be tuned by varying the size and spacing of the holes and to determine the best performance for the warning sound application a number of systems were tested in the ISVR anechoic chamber. The final design was then implemented and tested on a vehicle. The change in sound quality with direction can give a useful cue to pedestrians about the location of the vehicle.

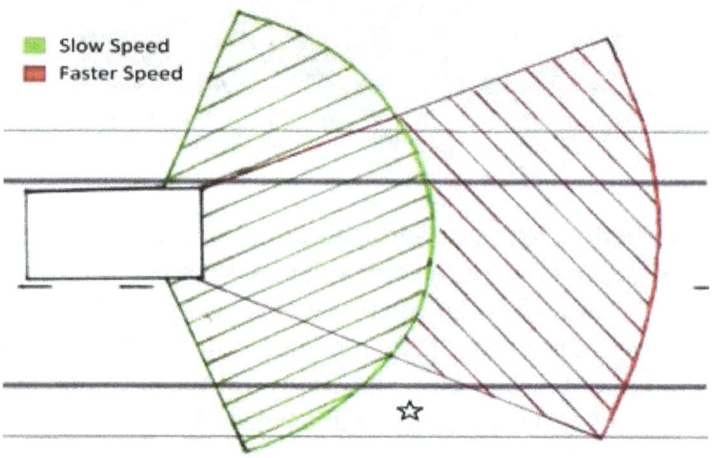

Required directivity

Specific Systems

Enhanced Vehicle Acoustics

Enhanced Vehicle Acoustics (EVA), a company based in Silicon Valley, California and founded by two Stanford students with the help of seed money from the National Federation of the Blind, developed an after market technology called "Vehicular Operations Sound Emitting Systems" (VOSES). The device makes hybrid electric vehicles sound more like conventional internal combustion engine cars when the vehicle goes into the silent electric mode (EV mode), but at a fraction of the sound level of most vehicles. At speeds higher than between 20 miles per hour (32 km/h) to 25 miles per hour (40 km/h) the sound system shuts off. The system also shuts off when the hybrid combustion engine kicks in.

VOSES uses miniature, all-weather audio speakers that are placed on the hybrid's wheel wells and emit specific sounds based on the direction the car is moving in order to minimize noise pollution and to maximize acoustic information for pedestrians. If the car is moving forward, the sounds are only projected in the forward direction; and if the car is turning left or right, the sound changes on the left or right appropriately. The company argues that *chirps, beeps and alarms are more distracting than useful*, and that the best sounds for alerting pedestrians are carlike, such as "*the soft purr of an engine or the slow roll of tires across pavement.*" One of the EVA's external sound systems was designed specifically for the Toyota Prius.

The Electric Citroën C1 with ECTunes external sound system.

ECTunes

ECTunes is developing a system that utilizes directional sound equipment to emit noise when and where it is needed. According to the company, its technology sends audible signals only in the direction of travel, thus allowing the vehicle to be heard by those who may be in the car's path, without disturbing others with unwelcome noise. Insero Horsens, a Danish venture company, has provided a significant investment to help ECTunes fully develop its technology.

The ECTunes system, and most others so far disclosed, use a control box, with software, digital amplifiers and weather-friendly external speakers. ECTunes' system connects to the car, and reads speed and acceleration, shutting down when the car reaches Cross-over speed as set by existing regulation as well as regulation under development such as Quiet Road Transport Vehicles (QRTV), at which point the tires and wind are making noise of their own. The company is currently selling products to OEM's, mainly small series production, and to the after market, and has also a new mass production unit in prototype stage

SoundRacer EVS

SoundRacer AB is a Swedish manufacturer of electronic engine sound devices established 2008 to develop sound technology for improved driving experience in internal combustion engine (ICE) cars. In 2012 the company released their first version of an Electric Vehicle Electronic Engine Sound System, based on the same technology.

SoundRacer EVS use only real engine sound recordings as the base for the sounds, following the legislation guidelines that the sound "should sound similar to the sound of a vehicle of the same category equipped with an internal combustion engine". As a result, the sounds will also be less likely to be disturbing compared to adding new types of synthetic sounds into the traffic environment.

The sound modules are developed to give a realistic ICE sound impression indicating if the vehicle is idling, accelerating, decelerating or cruising. The different sound characters and sound volume is determined from information about the actual speed of the vehicle and rate of speed change.

The company targets two different applications with the products: 1. EV warning sounds, to make electric vehicles meet the upcoming legislations, and 2. Increased driving experience to make electric vehicles more popular among customer groups that appreciate real engine sounds.

Many manufacturers of electric vehicles like cars, motorcycles and scooters are now in the process of evaluating sound systems for their products. To assist them in the prototyping process the SoundRacer EVS modules have functions that will give EV manufacturers the tools to test different sounds and settings. Several files with sounds and parameter settings, such as the speed for muting the sound, sound levels and simulated gear changing, can be created with a computer program and stored on a MicroSD card. Changing sounds and settings during test drives is then performed with a touch of a button.

SoundRacer AB delivers complete systems with recorded engine sounds from cars and motorcycles, including V8, V12 and V-Twin, CAN bus, analog and digital speed/rpm input and weather protected, IP68, speakers.

Fisker Automotive

The Fisker Karma has an automatic warning sound-generator.

The 2011 Chevrolet Volt includes a manually activated warning sound.

Fisker Automotive developed a sound-generator that was incorporated in its Fisker Karma luxury plug-in hybrid electric vehicle, released in 2011. According to the carmaker, the sound is designed to both alert pedestrians and enhance the driver experience, and the warning noise will be emitted automatically. The Fisker Karma emits a sound through a pair of external speakers embedded in the bumper. According to a company spokesman the sound is a mix between a "*Formula One car and a starship*".

The developing process took between nine months to a year, and three sound companies sent in synthesized WAV file samples that were evaluated by Fisker employees and executives. The prospective sounds were studied in an audio chamber to allow engineers to evaluate the sounds without other noise interfering. After testing the candidate sounds in different locations relative to the vehicle, Fisker fine-tuned the final sound with its own equipment. The warning sound is activated when the car is traveling at less than 25 mph (40 km/h).

Ford

The 2012 Ford Focus Electric was planned to include warning sounds for pedestrians. Ford Motor Company developed four alternative sounds, and in June 2011 involved the electric car fans by asking them to pick their favorite from the four potential warning sounds through the Focus Electric Facebook page. However, ultimately Ford decided to hold off including warning sounds unless federal legislation required it, and no such system was implemented on the production vehicle.

General Motors

General Motors' first commercially available plug-in hybrid electric vehicle, the Chevrolet Volt, introduced in December 2010, includes warning sounds for pedestrians. GM's system is called Pedestrian-Friendly Alert System and it is manually activated by the driver, but future generations probably will include an active system. The automaker conducted a test with a group of the visually challenged at Milford Proving Grounds in order to evaluate the audible warning systems on the Volt when a pedestrian is in the car's proximity. The system uses the car's horn to emit a series of warning chirps, like a low tone of a horn, enough to provide an alert but not to startle. According to GM engineers, the biggest challenge is "developing an active system that can distinguish a pedestrian from another vehicle"; otherwise, the sound will go off frequently, producing noise pollution instead.

Hyundai

Hyundai developed a warning noise for called the Virtual Engine Sound System (VESS). The system, which was introduced in September 2010 on its test fleet of BlueOn electric hatchbacks, provides synthetic audio feedback mimicking the sound of an idling internal combustion engine.

The 2011 Hyundai Sonata Hybrid is the first mass production car manufactured by Hyundai to include the warning sound system. In 2010 the carmaker decided to have a button on the Sonata Hybrid's instrument panel to turn the VESS on and off, but after the enactment of the Pedestrian Safety Enhancement Act of 2010, signed into law by President Obama in early 2011, and learning that the U.S. National Highway Traffic Safety Administration would not allow such switches to avoid the noise device to be turned off, Hyundai decided not to install the button, and the first Sonata Hybrids destined for the U.S. market had to be altered to remove the switch.

Lotus Engineering

Lotus Engineering, a consultancy group of British sports carmaker Lotus Cars, partnered in 2009 with Harman Becker, a producer of audio systems, to develop and commercialize a synthetic automotive audio systems. Lotus has worked on a number of hybrid and electric vehicles and its engineers thought they would be safer if these vehicles made a noise while moving around the factory. Originally developed to cancel out intrusive noises inside a car, the noise canceling system was adapted so that it could also simulate engine sounds that change with speed and use of the throttle, providing audible "feedback" to drivers of vehicles with a silent engine. At the same time, and through the addition of external speakers, the sound system allows pedestrians to hear the noise too, but optionally there can be a different sound within the car from the one that is emitted for the outside. Lotus used a Toyota Prius to demonstrate the device but did not reveal if it intended to bring this technology to market.

Lotus' synthetic sound system was incorporated in the Lotus Evora 414E Hybrid, a concept plug-in hybrid unveiled at the 2010 Geneva Motor Show. The system, called HALOsonic Internal and External Electronic Sound Synthesis, is a suite of noise solutions that uses patented technologies from Lotus and Harman International. The audio system generates engine sounds inside the vehicle through the audio system. The system also generates the external sound through speakers mounted at the front and rear to provide a warning to increase pedestrian safety. The system comes with four driver-selectable engine sounds, two of which have been designed to have characteristics of a multi-cylinder conventional V6 and V12 engine.

Nissan

2011 Nissan Leaf equipped with warning sound pause switch.

Vehicle Sound for Pedestrians or VSP is a Nissan-developed warning sound system in electric vehicles. The Nissan Leaf was the first car manufactured by Nissan to include VSP, and the electric car includes one sound for forward motion and another for reverse. The VSP was also used in the Nissan Fuga hybrid launched in 2011. The system developed makes a noise easy to hear for those outside to be aware of the vehicle approaching, but the warning sounds do not distract the car occupants inside. Nissan explained that during the development of the sound they studied behavioral research of the visually impaired and worked with cognitive and acoustic psychologists, including the National Federation of the Blind, the Detroit Institute of Ophthalmology, experts from the Vanderbilt University Medical Center and a Hollywood sound design studio.

Nissan's Vehicle Sound for Pedestrians is a sine-wave sound system that sweeps from 2.5 kHz at the high end to a low of 600 Hz, a range that is easily audible across age groups. Depending on the speed and whether the Leaf is accelerating or decelerating, the sound system will make sweeping, high-low sounds. For example, when the Leaf is started the sound will be louder, and when the car is in reverse, the system will generate an intermittent sound. The sound system ceases operation when the Nissan Leaf reaches 30 km/h (18.6 mph) and engages again as the car slows to under 25 kilometres per hour (16 mph). For the 2011 Leaf, the driver could turn off sounds temporarily through a switch inside the vehicle, but the system automatically reset to "On" at the next ignition cycle. The system is controlled through a computer and synthesizer in the dash panel, and the sound is delivered through a speaker in the front driver's side wheel well. Nissan said that there were six or seven finalist sounds, and that sound testing included driving cars emitting various sounds past testers standing on street corners, who indicated when they first heard the approaching car. Nissan removed the ability to disable the pedestrian alert between model year 2011 and 2012 in anticipation of the U.S. ruling to be issued by the National Highway Traffic Safety Administration.

The Leaf's electric warning sound had to be removed for cars delivered in the U.K., as the country's law mandates that any hazard warning sound must be capable of being disabled between 11:00 pm and 6:00 am, and the Leaf's audible warning system does not allow for such temporary deactivation. For the 2014 UK model of the car, the VSP system is enabled by default, though a button on the dash permits drivers to disable the system until the next time the car is switched on.

Toyota

Since August 2010, a warning device for retrofitting the Toyota Prius is available for Japanese consumers.

Toyota Motor Company teamed up with Fujitsu Ten to develop an automatic warning system for hybrids and electric vehicles to alert pedestrians when the car is propelled by its electric motor. The companies also studied the development of a system that would change the alarm's tune and volume with the assistance of an obstacle-detection radar.

On August 2010 Toyota began sales of an onboard device designed to automatically emit a synthesized sound of an electric motor when the Prius is operating as an electric vehicle at speeds up to approximately 25 kilometres per hour (16 mph). The device will be available in Japan through authorized Toyota dealers and Toyota genuine parts & accessories distributors for retrofitting on the third-generation Prius at a price of ¥12,600 (~US$150) including the consumption tax. The alert sound rises and falls in pitch according to the vehicle's speed, thus helping indicate the vehicle's proximity and movement to nearby pedestrians. Toyota is planning to use other versions of the device for use in gasoline-electric hybrids, plug-in hybrids, electric vehicles as well as fuel-cell hybrid vehicles planned for mass production. The device meets the 2010 government regulations issued for hybrid and other near-silent vehicles.

Toyota's Vehicle Proximity Notification System (VPNS) was introduced in the United States in all 2012 model year Prius family vehicles, including the Prius v, Prius Plug-in Hybrid and the standard Prius. The system is being introduced to comply with the Pedestrian Safety Enhancement Act of 2010.

Volkswagen

Volkswagen offers a so-called "e-Sound" module on its electric and hybrid vehicles such as the e-Up, e-Golf and the GTE hybrid range. It provides a pedestrian warning sound up to 30 km/h.

Other Carmakers

Tesla Motors and Think Global, both manufactures of electric cars already in the market, are assessing this safety issue. Ford Motor Company is developing a system for emitting external sounds to future hybrids and electrics, including its Focus BEV, scheduled for 2011, and a next-generation hybrid and plug-in hybrid vehicle planned for 2012. Nancy Gioia, Ford's Director for Global Electrification commented that "*car companies should consider standardizing tones from future hybrids and electrics to avoid a cacophony of confusion on the streets.*"

Open Charge Point Protocol

The Open Charge Point Protocol (OCPP) is a Simple Object Access Protocol (SOAP)-based protocol for communication between an EV Charging Station and a central Back Office system. OCPP is an international open standard which is available for free, no license fees are required. Developed in 2009 for the EV infrastructure market, it is supported today by many large players in the EV industry such as utilities, Charging Station manufacturers, and back-office software suppliers. As such, OCPP is designed to be vendor independent, thereby providing the opportunity for infrastructure operators in choosing EV Charging Stations, and for vendors to supply EV Charging Stations to any infrastructure operator.

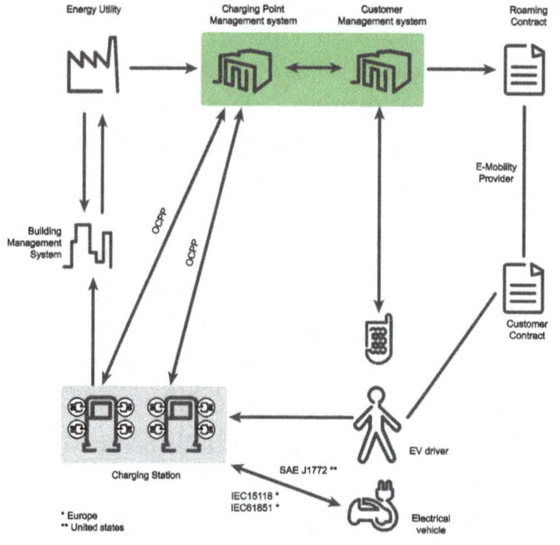

SOAP is a framework which makes it possible to send messages between components over the internet. The advantage of SOAP is that the facilities for sending and receiving messages are covered by the standard. This makes possible rapid implementation.

The content of a SOAP message is drawn up according to the Extensible Markup Language (XML) standard. This language is related to the HTML that we know from the Internet. In addition to written text, XML messages may also contain images and executable code. The big advantage is that messages are sent in legible text.

Implementation

EVlink Charging Stations support OCPP 1.5 with software packages:

- Webserver 2.7.1 or higher,
- EVSE 2.7.1 or higher,
- Component 1.0.6. or higher.

Settings and data transmission between the Charging Stations and the IT Backend should be checked and tested before any site implementation.

You should contact Schneider Electric prior to your project definition to check if the planned implementation is validated.

For Charging Stations with earlier software versions, the Charging Station software must be upgraded following EVlink commissioning Guide instructions.

Each IT Backend integration requires some specific parameters to be setup at commissioning. This configuration step can be performed manually through the commissioning tool.

Schneider Electric recommends to upload to the Charging Station a ready to use setting file for each IT Backend tested by Schneider Electric. This file shall be requested to Schneider Electric for each customer and each Charging Station location.

Supported Functions

25 operations are described within OCPP 1.5. 10 of these are initiated by the Charging Station and 15 by the IT Backend.

When communication takes place between a Charging Station and the IT Backend it always begins with a request. In the OCPP specification this is described with operationname.req(). The recipient of the message will always reply with a confirmation. In the OCPP specification a confirmation is described as operationname.conf().

The figure on the opposit page shows the communication flow for the operation Authorize in a charging session context. The Charging Station sends a message to the IT Backend Authorization. req() and gets an answer in the form of Authorization.conf(). For the sake of clarity the message names and the rest of the communication flow are not displayed in this figure.

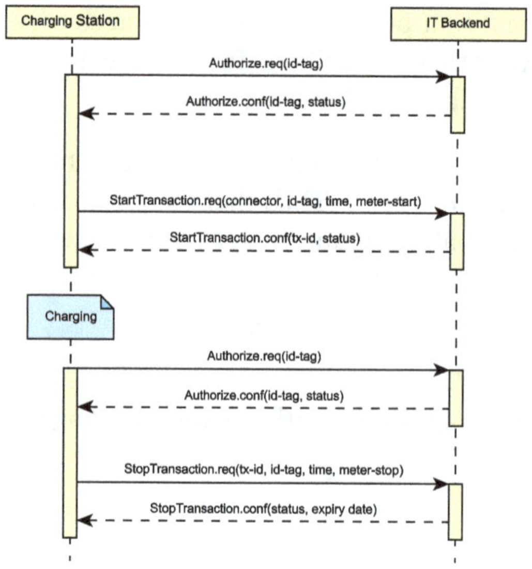

Functions Initiated by a Charging Station

The general description of these functions is in the following document "Open Charge Point Protocol Interface description between Charge Point and Central System".

The following functions are supported by the Charging Station:

Starting up a Charging Station (Boot Notification)

Once a Charging Station has been turned on and started it will attempt to make contact with the IT Backend. The Charging Station sends a message containing details of the configuration, brand and type to the IT Backend.

The IT Backend will check if the Charging Station is known in the system. If this is the case the IT Backend sends a reply stating whether or not the Charging Station is accepted. If the Charging Station is accepted the answer will also include the time and date of the IT Backend and the set heartbeat interval.

The table below is provided as an example:

Field	Sample Value	Optional
chargePointVendor	Schneider Electric	
chargePointModel	MONOBLOCK	
chargePointSerialNumber	3N134420451C1S1B7551700007	✓
chargeBoxSerialNumber	EV.1S22P333N13483040030015006582	✓
firmwareVersion	2.4.76	✓
iccid	89314404000056919443	✓
imsi	204043728031302	✓

If the Charging Station is rejected, it will go to error mode.

If the Charging Station is unable to contact the IT Backend, or receives a message stating that it is not accepted, it will regularly continue attempting to make contact.

Heartbeat

Once the Charging Station has received confirmation from the IT Backend that it is connected, it will send a heartbeat to the IT Backend at an interval which was designated by the IT Backend in the Boot-Notification message response. This informs the IT Backend that the Charging Station is still alive. In response to a heartbeat the IT Backend always sends back the system's current time and date.

The Charging Station does not send a Heartbeat message if another message has been sent to the IT Backend during the configured heartbeat interval.

Authorize

The authorization process enables the user to start a charge on the Charging Point and unlocks a connector if allowed to.

Note: stopping a charge transaction only needs clearance from the Central System if the user stopping the charge session is different than the one who started it.

StartTransaction

This message is sent as soon as a charge is started.

The transaction starts even if the Charging Station is offline and can not reach the IT Backend. In this case, the StartTransaction Request will be sent to the Supervision as soon as the Charging Station comes back online.

If Charging Station does not have a RFID reader, the idTag which is sent in StartTransaction message is the one used in the OCPP configuration file (default Id Tag field).

MeterValues

A Charging Station may sample the electricity meter or other sensor/transducer hardware to provide extra information about its meter values.

It is up to the Charging Station to decide when it will send meter values.

This can be configured using the Change Configuration message.

StopTransaction

If the user wants to stop his/her charge, he/she needs to:

Step	Action
1	Identify himself/herself using the charge pass at the Charging Point.
2	The Charging Point sends a request to the IT Backend to stop charging. This request includes the final meter reading, transaction number, pass ID and the current time.
3	The IT Backend will indicate whether the user is authorized.

StatusNotification

A Charging Point sends a notification to the Central System to inform the Central System about a status or error condition within the Charging Point.

Dealing with Errors

When an error is detected by a Charging Point, it will relate the error to the IT Backend. As soon as the Charging Point enters in Error Mode, it sends a status notification message with following information:

Field	Sample Value	Optional
ConnectorScope	1 (or 2)	
status	Faulted	
errorCode	Over Current Failure	
vendorErrorCode	21	✓

Vendor Error Code contains Schneider Electric information about the current error. Following table describes the error according to error code:

Id	Error description
1	Lost communication with RFID reader
2	Lost communication with display
3	Cannot connect to master board
4	Incorrect plug lock state
5	Incorrect contactor state
6	Incorrect surge arrestor state
7	Incorrect anti intrusion state
8	Cannot connect to US daughter board
9	Configuration file missing, corrupted or already open
10	Incorrect shutter lock state
11	Incorrect circuit breaker state
12	Lost communication with power meter
13	Incorrect voltage level
14	Incorrect socket state
15	Incorrect charging phase number
16	Lost communication with cluster manager
17	Mode3 communication error (CP error)
18	Incorrect cable state (PP error)
19	Default EV charging cable disconnection
20	Short circuit FP1
21	Overcurrent
22	No energy available for charging

23	HighTemperature
24	WeakSignal
25	Tilt
26	TiltSensorFailure
27	Shock
28	ShockSensorFailure
29	ServiceDoorFailure
30	Supervision Error
31	Reserved
32	LowDiskSpace

DiagnosticsStatusNotification

Charging Station sends a notification to inform the IT Backend when a diagnostics upload has finished.

FirmwareStatusNotification

Charging Station sends notifications to inform the IT Backend about the progress of the firmware update.

Function Initiated at IT Backend

The following functions are supported by the IT Backend:

ClearCache

If the cache is enabled, Charging Station memorizes previously authorized cards in a cached list. The IT Backend can send a ClearCache message for clearing the Charging Station's cache.

ChangeAvailability

The Central System can request the Charging Station or the Charging Point to change its availability to inoperative or operative.

A Charging Point is considered available ("operative") when it is charging or ready for charging. A Charging Point is considered unavailable when it does not allow any charging.

ChangeConfiguration

ChangeConfiguration enables a IT Backend to modify different Charging Station settings.

These include the heartbeat interval.

GetConfiguration

The Central System issues this request to obtain the value of configuration settings from the Charging Station.

GetDiagnostics

The IT Backend can ask a Charging Station to send diagnostic data. After such a request the Charging Station will upload a MaintenanceReport.html file which contains the entire configuration settings and status.

When the file is uploaded the Charging Station sends a DiagnosticsStatusNotification message with the status of the upload.

The diagnostic file is uploaded via FTP. The FTP URL is of format: ftp://user:password@host:port/path in which the parts user: password@,:password or: port may be excluded.

Note: user should not contain any of the following: ":", "@" password should not contain "@".

RemoteStopTransaction

IT Backend can request a Charging Station to stop a transaction by sending a RemoteStopTransaction message to Charging Station with the identifier of the transaction.

Charging Station replies with RemoteStopTransaction response message to indicate whether it is indeed able to stop the transaction.

In case of a Charging Station allowed authentification, the connector is still locked after the stop transaction. Active user has to use his pass to unlock the connector if the charge has been started locally or the IT Backend has to send an Unlock Connector message.

Reset

The Central System shall send this request to the Charging Station to reset it. The Central System can request a hard or a soft reset.

In both cases the transaction in progress will be terminated.

In case of a Reset Hard, the Charging Station becomes unavailable and then restarts. If a charge was in progress at this time, the plug is locked after the Charging Station has finished booting. Reset Soft does not restart the Charging Station.

UnlockConnector

IT Backend can unlock a connector of a Charging Point with the UnlockConnector Message.

The Charging Point replies with an UnlockConnector response message which indicates whether the Charging Station is able to unlock its connector.

In case of a transaction in progress on the specific connector, then the transaction is stopped before unlocking the connector.

UpdateFirmware

When a firmware update is ready the IT Backend sends a message to the Charging Station containing the location where the update can be downloaded. The Charging Station then sends a confirmation to the IT Backend that the message has been received. Then the Charging Station will download and install the update. During the installation process the Charging Station informs the IT Backend of the current status, whether the update has been successfully downloaded or not and if the installation has succeeded.

Benefits of OCPP

First, because OCPP is a standardized protocol used by many charging station vendors, you have options. As mentioned before, if your charging station vendor goes out of business, and their stations are OCPP-compliant, you can connect to another OCPP-compliant network using your original charging stations. As another example, if you dislike your current vendor because of the fees or the service they provide, you can just connect your OCPP-compliant stations to a new OCPP-compliant network.

Second, because it's so easy for a station owner to switch vendors, we have to make sure our charging stations and the Sema Connect network are the best that they can possibly be. We constantly innovate so you can get more features that bring you value. You have the freedom to switch to a different network if you have to.

Third, OCPP allows you to mix and match stations and networks from different vendors. If you're thinking of expanding your stations at your property but prefer the network capabilities from a different vendor than your original one, you have the freedom to combine your old stations with new stations on the new vendor's network and easily manage your mix of stations through one portal.

OCPP Benefits to the Drivers

OCPP benefits drivers in two ways:

First, they reap the benefits of you using the best charging stations and network. When you have the freedom to switch to the vendor with the best features and management tools, they get to use stations that will give them the best charging experiences.

Second, the entire goal of OCPP is to make it possible to build a network of EV charging stations. Part of the reason why EV adoption has taken off in Europe faster than in the US is that there is more standardization across the EVSE industry. Widespread adoption of OCPP leads to interoperability between devices and networks from different brands, which leads to the freedom to switch to the best network for a property's needs, which leads to even more places to charge your car!

Drivers want to be able to charge their vehicles everywhere they go, whether it's their home, work, or vacation spots. They expect to find premium charging experiences from stations that don't use proprietary protocols and proprietary payment systems.

This OCPP security specification has been designed to meet a number of security objectives:

- To allow the creation of a secure communication channel between the Central System and Charge Point. The integrity and confidentiality of messages on this channel should be protected with strong cryptographic measures.

- To provide mutual authentication between the Charge Point and the Central System. Both parties should be able to identify who they are communicating with.
- To provide a secure firmware update process by allowing the Charge Point to check the source and the integrity of firmware images, and by allowing nonrepudiation of these images.
- To allow logging of security events to facilitate monitoring the security of the smart charging system.

A basic security profile is included in this specification that does not meet the first two requirements. It is provided for systems that do not require security (e.g. lab installations or test setups). For field operation it is highly recommended to use a security profile with TLS.

Design Considerations

The security specification was designed to fit into the approach taken in OCPP. Standard web technologies whenever possible to allow cost-effective implementations using available web libraries and software. The security specification is designed to support both the SOAP (OCCP-S) and JSON (OCPP-J) variants of OCPP. Only security measures available for both technologies have been used. No application layer security measures are included. Based on these considerations, the security specification is based on TLS and public key cryptography using X.509 certificates. Because the Central System usually acts as the server, different users or role-based access control on the Charge Point are not implemented in this standard. To mitigate this, it is recommended to implement access control on the Central System. To make sure the mechanisms implemented there cannot be bypassed, OCPP should not be used by engineers performing maintenance to Charge Points locally at the Charge Point.

Password Management

If HTTP Basic authentication is used to authenticate the Charge Point, the Charge Point requires a password. The password SHALL be stored as an option under the key AuthorizationKey.

To set a charge point's authorization key via OCPP, the Central System SHALL send the Charge Point a ChangeConfiguration.req message with the key AuthorizationKey (which is write-only) and as the value a 40-character hexadecimal representation of the 20-byte authorization key. If the Charge Point responds to this ChangeConfiguration.req with a ChangeConfiguration.conf with status Accepted, the Central System SHALL assume that the authorization key change was successful, and no longer accept the credentials previously used by the charge point.

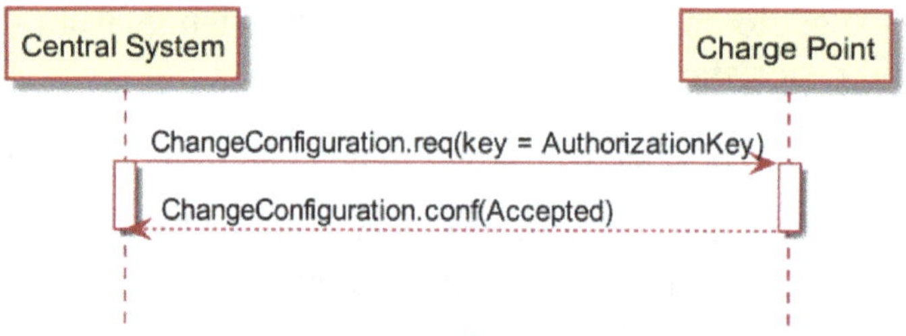

Figure: Password Management (happy flow)

If the Charge Point responds to the ChangeConfiguration.req with a ChangeConfiguration.conf with status Rejected or NotSupported, the Central System SHALL keep accepting the old credentials. While the Central System SHALL still accept a connection from the Charge Point in this case, it MAY treat the Charge Point's OCPP messages differently, e.g. by not accepting the Charge Point's boot notifications.

Different user names and passwords SHOULD be used for different Charge Points. Passwords SHOULD be generated randomly to ensure that the passwords have sufficient entropy. Furthermore, the Central System SHOULD only store salted password hashes, not the passwords themselves. Similarly, the Central System SHOULD NOT put the passwords in clear-text in log files or debug information. In this way, if the Central System is compromised not all Charge Point password will be immediately compromised.

On the Charge Point the password needs to be stored in clear-text. Extra care SHOULD be taken into storing it securely. Definitions of mechanisms how to securely store the credentials are however not in scope of the OCPP Security Profiles.

Key Management

The OCPP standard uses a number of public private key pairs for its security. To manage these keys, methods to update and revoke the keys must be provided. Only the methods to update the keys on the Charge Point are considered here. Updating keys on the Central System or at the manufacturer has to be done with other methods than OCPP. If TLS with Client Side certificates is used, the Charge Point requires a "Charge Point certificate" for authentication against the Central System. A unique Charge Point certificate SHOULD be used for each Charge Point.

Table: Keys used in the OCPP security specification.

Key	Private Key Stored At	Description
Central System certificate	Central System	Certificate used to authenticate the Central System
Charge Point certificate	Charge Point	Certificate used to authenticate the Charge Point
Firmware Signing Certificate	Manufacturer	Certificate used to verify the firmware signature

Certificate Properties

All certificates shall use a private key that provides security equivalent to a symmetric key of at least 112 bits. This is the key size that NIST recommends for the period 2011-2030. For RSA or DSA this translates into a key of at least 2048 bits long. For elliptic curve cryptography it translates into a key of at least 224 bits long.

For all cryptographic operations only the algorithms recommended by ENISA in as suitable for use in future systems SHALL be used. This restriction includes the signing of certificates in the certificate hierarchy. For signing by the certificate authority RSA-PSS, or ECDSA SHOULD be used, for computing hash values the SHA256 algorithm SHOULD be used.

The certificates shall be stored and transmitted in the X.509 format encoded in the Privacy-enhanced Electronic Mail (pem) format. All certificates shall include a serial number. The subject field of the certificate SHALL contain the organization name of the certificate owner in the organizationName RDN.

For the Central System certificate, the subject field shall contain the URL or IP address of the server in the commonName RDN.

For the Charge Point certificate, the subject field shall contain a commonName RDN that allows the Charge Point to be uniquely identified. For instance, the commonName could be set to a serial number uniquely identifying the Charge Point.

For all certificates the X.509 Key Usage or Extended Key Usage extension should be used to restrict the usage of the certificate to the operations for which it will be used. For instance, these extensions can be used to indicate that the Firmware Signing certificate is only to be used for signatures, not encryption.

Certificate Hierarchy

The OCPP protocol supports the use of two separate certificate hierarchies:

- The Charge Point Operator hierarchy that contains the Central System, and Charge Point certificates.
- The Manufacturer hierarchy that contains the Firmware Signing certificate.

The Charge Point Operator MAY acts as a certificate authority for the Charge Point Operator hierarchy. This could give the advantage that the online verification of Charge Point client side certificates can be done within the Charge Point Operators networks, simplifying the network architecture. The root certificates containing the private keys MUST however be well protected.

As the Manufacturer is usually a separate organization from the Charge Point Operator, a trusted third party SHOULD be used as a certificate authority. This is essential to have non-repudiation of firmware images.

The Central System can update the root certificates stored on the Charge Point using the *UpdateRootCertificate.req* message. The new root certificate MUST be signed with the old root certificate, and the Charge Point MUST check the validity of the new certificate using this signature.

Updating the Charge Point Certificates

To facilitate the management of the Charge Point client side certificate, a key update procedure is provided. The key update procedure is triggered in two cases:

- The Central System requests the Charge Point to update its key using the *UpdateChargePointCertificate.req* message.
- The Charge Point detects that the certificate it is using will expire in one month.

A key update SHOULD be performed after installation of the Charge Point, to change the key from the one initially provisioned by the manufacturer (possibly a default key).

When the key update process is triggered, the Charge Point SHALL generate a new public / private key pair using one of the key generation functions. The Charge Point SHALL send the public key in form of a Certificate Signing Request (CSR) as described in the X.509 standard

The Central Server SHOULD not sign the certificate itself, but instead forwards the CSR to a dedicated certificate authority server managing the certificates for the Charge Point infrastructure. The certificate authority SHOULD implement strong measures to keep the certificate signing private keys secure. Furthermore, the private key generated by the Charge Point during the key update process SHALL NOT leave the Charge Point at any time, and SHALL NOT be readable via OCPP.

The Charge Point SHALL verify the validity of the signed certificate in the SignCertificate.conf message, checking at least the period when the certificate is valid, the properties and that it is part of the Charge Point Operator certificate hierarchy. If the certificate is not valid, the Charge Point SHALL discard the certificate, and trigger a 'Invalid Charge Point certificate' security event.

The Charge Point SHALL switch to the new certificate as soon as it is valid, that is, the current date and is after the 'Not valid before' field in the certificate. If the Charge Point contains more than one valid certificate of the same type, it SHALL use the newest certificate, as measured by the start of the validity period. The Charge Point MAY discard a certificate one month after its validity has expired.

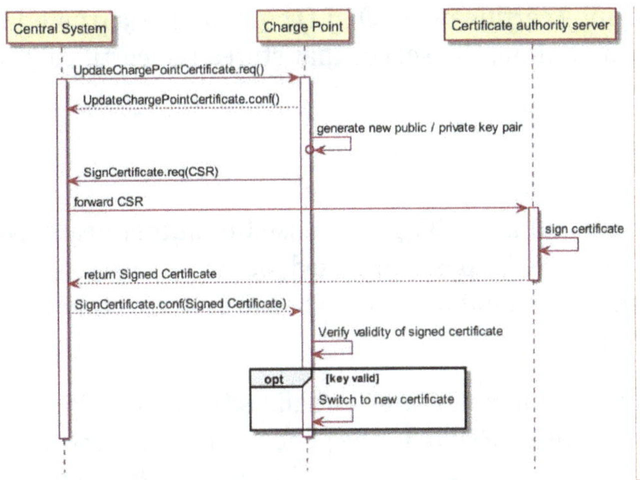

Certificate Revocation

Different methods are recommended for certificate revocation.

Table: Recommended revocation methods for the different keys.

Key	Revocation
Central System certificate	Fast expiration
Charge Point certificate	Online verification
Firmware Signing certificate	Online verification

Fast expiration should be used to revoke the Central System certificate. With fast expiration, the certificate is only valid for a short period, less than 24 hours. After that the server needs to request a new certificate from the Certificate Authority. In this way, if the certificate is compromised, the impact is reduced to only a short period. Also, the Charge Points do not need to implement revocation lists or online certificate verification. This simplifies the implementation of certificate management at the Charge Point.

The Central System should use online certificate verification to verify the validity of the Charge Point certificates. It was recommended that a separate certificate authority server is used to manage the certificates. This server should also keep track of which certificates have been revoked. The Central Systems then verifies the validity of the certificate with the certificate authority server. This allows for immediate revocation of Charge Point certificates. Revocation of Charge Point certificates will happen for instance when a Charge Point is removed. This is more common than revoking the Central System certificate, which is normally only done when it is compromised.

Online verification for the Firmware Signing certificate should be performed. As part of the firmware updated process, the Charge Point downloads the Firmware Signing certificate whenever it performs a firmware update. Prior to providing the certificate for firmware validation to the Charge Point, the Central system should validate both, the certificate and the signed firmware update.The Firmware Signing certificate SHOULD be signed by a trusted third party. This means that to verify the certificate online, the server that stores the certificate may need (indirect) internet access.

Installation

As stated above, unique credentials SHOULD be used to authenticate each Charge Point to the central system, whether they are the password used for HTTP Basic Authentication or the Charge Point certificate. These unique credentials have to be put on the Charge Point at some point during manufacturing or installation.

It is RECOMMENDED that the manufacturer initializes the Charge Point with unique credentials during production. The credentials SHOULD be generated using a cryptographic random number generator, and installed in a secure environment. They SHOULD be sent to the CSO over a secure channel, so that the CSO can import them in the central system.

If Charge Point certificates are used, the manufacturer MAY sign these using their own certificate.

It is RECOMMENDED that the CSO immediately updates the credentials after installation. Before the credentials have been updated, the Central System may restrict the functionality that the Charge Point can use.

Security Profiles

This topic defines the communication security setting for OCPP. To allow for different situations and needs for security, three security profiles are defined:

- Basic security profile.

- TLS with Basic Authentication.
- TLS with Client Side Certificates.

To choose the right profile in each situation, the following factors SHOULD be considered:

- The basic security profile does not include authentication for the Central System, or measures to set up a secure communication channel. Therefore, it should only be used in trusted networks, for instance in networks where there is a VPN between the Central System and the Charge Point.
- The Basic Security and TLS with Basic Authentication profiles both assume that the Central System acts as the server. This is always the case for OCPP-J (WebSockets). It is usually the case in OCPP-S (SOAP). For OCPP-S it is however allowed that the Charge Point acts as a server. If this functionality is desired, the TLS with Client Side certificates profile should be used.

For each profile the choices made for authentication and communication security are described below.

Table: Summary of OCPP Security Profiles.

Profile	Charge Point Authentication	Central System Authentication	Communication Security
Basic Security	HTTP Basic Authentication	None	None
TLS with Basic Authentication	HTTP Basic Authentication	TLS authentication using certificate	Transport Layer Security (TLS)
TLS with Client Side Certificates	TLS authentication using certificate	TLS authentication using certificate	Transport Layer Security (TLS)

The Charge Point and Central System shall only support one security profile at a time. If the Charge Point tries to connect with a different profile than the Central System is using, then the Central System shall terminate the connection, and vice versa. The security profile must be configured before OCPP communication is possible. Changing the profile that is used is therefore not part of the OCPP specification, and must be done through another method where Charge Points with different security profiles or different versions of OCPP are in use, it is possible to operate the Charge Points via connecting them to different addresses or ports of the Central System. For instance, the Central System server may have one TCP port for TLS with Basic Authentication, and another port for TLS with Client Side Certificates. From the perspective of this communication standard there is only one security profile in use per Central System. Using this construction, it is not necessary to add an extra handshake to the OCPP protocol to negotiate the security profile version.

Basic Security Profile

The Basic Security profile provides a minimum level of security. Charge Point authentication is done through a username and password. No measures are included to secure the communication channel.

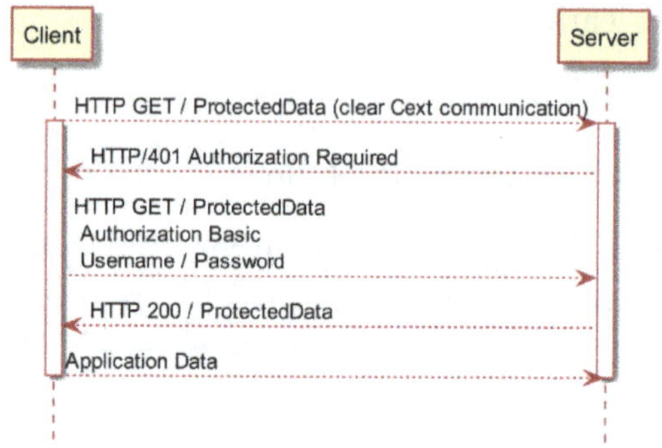

Figure: Basic Security Profile

Charge Point Authentication

The Charge Point SHALL authenticate itself to the Central System using HTTP Basic authentication. When using HTTP Basic authentication, the client, i.e. the Charge Point, has to provide a username and password with its request. For OCPP-J, the username should be equal to the charge point identity, which is the identifying string of the charge point as it uses it in the OCPP-J connection URL. The password should be a 20-byte key stored in the *AuthorizationKey* property on the charge point.With HTTP Basic, the username and password are transmitted in clear text, encoded in base64 only. Hence, this mechanism should only be used over connections that are already secured with other means. Basic authentication is used for compatibility with WebSockets.

Central System Authentication

In this profile, the Central System does not authenticate itself to the Charge Point. The Charge Points has to trust that the server is connects to is indeed the Central System.

Communication Security

No communication security measures are included in the profile.

TLS with Basic Authentication

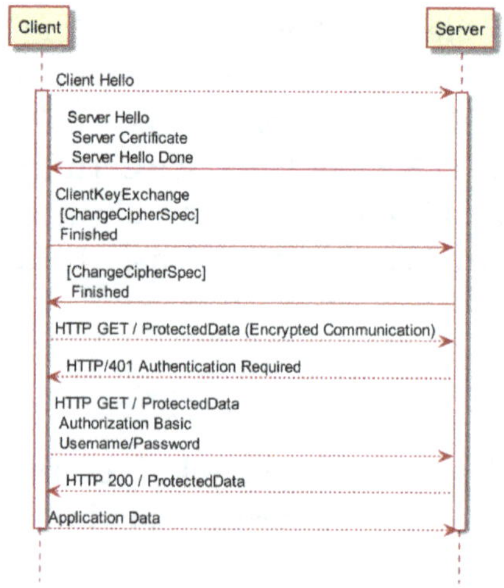

Figure: TLS with Basic Authentication

In the TLS with Basic Authentication profile the communication channel is secured using Transport Layer Security (TLS). The Central System authenticates itself using a TLS server certificate. The Charge Points authenticate themselves using HTTP Basic Authentication.

Charge Point Authentication

The Charge Point SHALL authenticate itself to the Central System using HTTP Basic authentication. Because TLS is used in this profile, the password will be sent encrypted, reducing the risks of using this authentication method.

Central System Authentication

The Central System acts as the TLS server, and will authenticate itself by using the Central System certificate as server side certificate. The Charge Point shall verify the certification path of the Central System's certificate according to the path validation rules. The Charge Point shall verify that the commonName includes the Central System's URL or IP address.

The Charge Point shall terminate the connection if the Central System does not own a valid or any certificate, of if the certification path is invalid. In that case the Charge Point shall trigger an 'Invalid Central System certificate' security event.

Regenerative Braking

Regenerative braking systems (RBSs) are a type of kinetic energy recovery system that transfers the kinetic energy of an object in motion into potential or stored energy to slow the vehicle down, and as a result increases fuel efficiency. These systems are also called kinetic energy recovery systems. There are multiple methods of energy conversion in RBSs including spring, flywheel, electromagnetic and hydraulic. More recently, an electromagnetic-flywheel hybrid RBS has emerged as well. Each type of RBS utilizes a different energy conversion or storage method, giving varying efficiency and applications for each type.

RBSs are installed along the drive train or fitted to the drive wheels of a vehicle where they inhibit the motion of the wheels using magnetic fields or mechanical torque. These methods of motion inhibition allow energy to be generated under braking, as opposed to friction brakes which simply waste away energy to slow the vehicle by turning the kinetic energy into thermal energy. Due to the maximum charging rate of the energy storage mechanisms, the braking force from a RBS is limited. Therefore, a traditional friction brake system is required to maintain the safe operation of a vehicle when heavy braking is necessary. RBS can improve fuel consumption and reduce the overall braking load taken on by the vehicles friction brakes, reducing the wear on the brake pads.

RBSs are used in almost every electric vehicles and hybrid electric vehicles. In addition, public transportation such as buses and bullet trains make use of RBSs to decrease the environmental impacts of the transportation fleet and save money.

Effectiveness of Regenerative Braking

To evaluate regenerative braking, we really need to look at two different parameters, efficiency and effectiveness. Despite sounding similar, the two are quite different. Efficiency refers to how well regenerative braking captures 'lost' energy from braking. Does it waste a lot of energy as heat, or does it turn all of that kinetic energy back into stored energy? Effectiveness, on the other hand, refers to how large of an impact regenerative braking really makes. Does it measurably increase your range, or will you not notice much of a difference?

Efficiency

No machine can be 100% efficient (without breaking the laws of physics), as any transfer of energy will inevitably incur some loss as heat, light, noise, etc. Efficiency of the regenerative braking process varies across many vehicles, motors, batteries and controllers, but is often somewhere in the neighborhood of 60-70% efficient. Regen usually loses around 10-20% of the energy being captured, and then the car loses another 10-20% or so when converting that energy back into acceleration, according to Tesla. This is fairly standard across most electric vehicles including cars, trucks, electric bicycles, electric scooters, etc.

Keep in mind that this 70% does not mean that regenerative braking will give an 70% range increase. This isn't going to bump your range from 100 miles to 170 miles. This simply means that 70% of the kinetic energy lost during the act of braking can be turned back into acceleration later.

This is why only reporting the efficiency of the system doesn't really mean much. Someone could be very efficient when they work, but if they only work an hour a day, they probably aren't accomplishing a lot. What should interest us more is the effectiveness of regenerative braking.

Effectiveness

This is where things get really interesting. The effectiveness of regenerative braking is a measure of how much it can increase your range. Does it make your theoretical range 5% further? 50% further?

As you've probably already guessed, the effectiveness of regenerative braking varies significantly based on factors including driving conditions, terrain and vehicle size.

Driving conditions have a large impact. You'll see much better effectiveness for regenerative braking in stop-and-go city traffic than in highway commuting. This should make sense, as if you're repeatedly braking, you'll recapture a lot more energy than if you simply drive for hours without touching the brake pedal. Terrain also plays a large role here too, as uphill driving doesn't give you much chance for braking, but downhill driving will regenerate a much larger amount of energy due to the long braking periods. On long downhills, regenerative braking can be used nearly constantly to regulate speed while continuously charging the battery.

Vehicle size may be the largest factor in the effectiveness of regenerative braking for the simple reason that heavier vehicles have much more momentum and kinetic energy. Just like a big flywheel is more effective than a small flywheel, a four-wheel electric car has a lot more kinetic energy when in motion than an electric bicycle or scooter.

Data for comparison can be somewhat hard to come by, tesla vehicles show you the regenerative braking power, such as 60 kW during hard braking, but that doesn't answer the more interesting question. We want to know how much energy we are recapturing over a trip, not how strong our brakes are each time we mash the pedal.

Fortunately, a number of Tesla drivers have reported back energy contribution data using different data tracking apps. Model S drivers have reported recapturing as much as 32% of their total energy use while driving up and then back downhill. This would effectively increase a 100 mile car's range to 132 miles, for example. A Model S P85D owner reported approximately 28% energy recapture (forum in Danish) and still others have reported recapturing between 15-20% of their total kWh usage on average during normal trips.

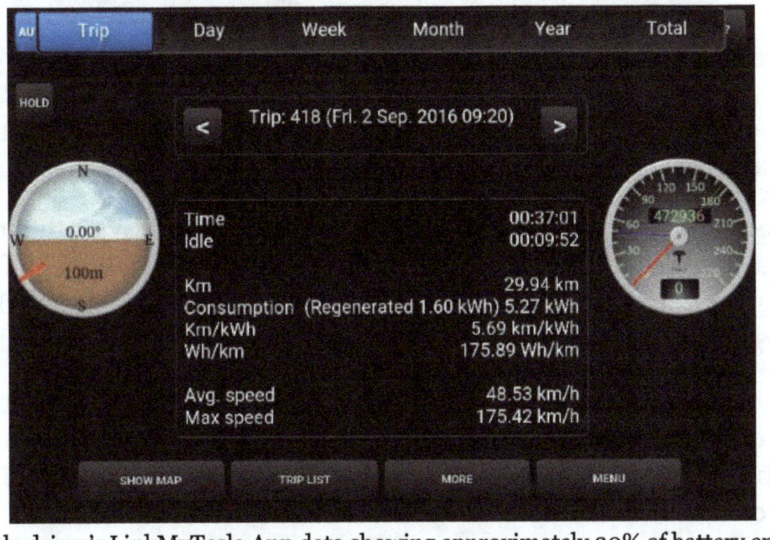

A Tesla driver's LinkMyTesla App data showing approximately 30% of battery energy usage recaptured using regenerative braking.

For smaller EVs such as personal electric vehicles, the numbers aren't quite as optimistic. On multiple electric bicycles with regenerative braking options, I've generally averaged around 4-5% regeneration, with a maximum of around 8% in hilly areas. Other personal electric vehicles including electric scooters and skateboards have similar results, usually in the lower single digits. Again, keep in mind this isn't the raw efficiency of the system (as in how much braking energy is lost in the energy transfer), it's the effectiveness (as in how much further your range increases due to the use of regenerative braking).

This is largely due to the lower weight of personal electric vehicles. They simply don't carry much momentum and thus have less kinetic energy to convert back into the battery.

Working Regenerative Braking

In the e-bike industry, regenerative braking can sometimes be used more as a marketing tool than as a feature. Because regenerative braking is generally only possible in electric bicycles with larger gearless motors, such e-bike manufacturers will tout the effectiveness of their models. At the same time, manufacturers of e-bikes with mid-drives and other geared motors that are incapable of regen braking will dismiss it as ineffective and simply not worth it.

Most mid-drive electric bicycles are not capable of regen braking

The truth of the matter is that for smaller and personal electric vehicles, regenerative braking isn't as effective as in electric cars, but it still has a multitude of advantages.

One of the biggest benefits of regenerative braking for smaller personal EVs is the added braking force. Some PEVs, such as the Xiaomi M365 electric scooter, only use regen braking for the front motor wheel, while relying on a traditional disc brake for the rear wheel. This means the scooter has two independent brakes with only one brake lever to activate them, reducing cost, weight and complexity.

Regenerative braking also allows the introduction of brakes to electric skateboards – a feat that was previously accomplished using the variable braking feature of your shoe sole on the pavement. With popular electric skateboards such as the Boosted Board reaching speeds of well over 20 mph, electric braking accomplished via regen is a much welcomed safety feature.

Another benefit of regenerative braking is extending the life of conventional brake parts such as cables and brake pads. These can be annoying to maintain and replace, especially since electric bicycles and scooters travel much farther and faster than their non-electric brethren and would otherwise wear out brake pads much quicker. One of my e-bikes has no regen due to having geared motors that freewheel, and it seems like I'm always tuning and adjusting the brakes. On regen-enabled e-bikes though, I find that I can often rely almost entirely on regenerative braking, meaning my brake pads see minimal use.

At the end of the day, regenerative braking will never be as effective in smaller vehicles as it is in larger ones simply due to physics. Because of this, the lack of regen in e-bikes and other PEVs isn't a deal breaker. However, the benefits of regenerative braking outside of simple energy recapture can't be ignored.

Traction Motor

A traction motor is an electric motor used to drive a piece of machinery. Consisting of heavy wire wound inside of the motor's case known as the field windings and wire wound around the motor's inner shaft called the armature, the electric power is transported between the two elements through brushes. The brushes are small, spring-loaded metallic components upon which the armature's exposed shaft rides. Many systems are used in the motor to redirect excess power as well as to use the electricity to reverse the motors and act as brakes.

There are many sizes and styles of traction motors in use around the world; however, the most commonly recognized form is found in a diesel locomotive. The diesel engine in a locomotive turns a large generator that supplies the power to the electric motor that actually propels the train. The official name for the motor that powers escalators, electric automobiles and even clothes-washing machines is a traction motor. Both alternating current (AC) and direct current (DC) electric motors have been used as traction motor designs on a vast array of machines, with the most successful design using the AC motor.

One reason for the use of a traction motor on a diesel locomotive is the ease with which the power can be applied to the wheels. Unlike an internal combustion engine, an electric motor begins making power in the form of torque, as soon as the motor receives electrical power. This allows the motor to begin easily pulling the load at very slow speeds. In the case of an internal combustion engine, the engine typically makes the majority of its pulling power in the mid to upper revolutions per minute (RPM) engine speed. With the electric motor coming on so strong at such a slow speed, the engineer is able to better control the starting of the train into motion without damaging the train, the cargo or the train's crew.

In an effort to reduce weight, gear boxes used to direct power from the motor to the wheels were replaced by placing the wheels directly onto the motor's center shaft. With the motor working as its own axle, the locomotive's chassis as well as the tracks were given a longer life due, in part, to the reduction of weight being applied to both components. By placing specially designed resistors in the motors, the speed control is much easier than the early traction motor models in which the engineer was continually forced to manually reduce the amount of current being sent to the motors in order to maintain a steady speed.

General Requirements of Traction Motors/Traction System

- High starting torque

  A traction motor must have high starting torque. For urban and sub urban lines, the acceleration (speed at starting) should be high.

- Simple Speed Control

  A traction motor speed control should be easy and simple such that motor can be easily start and stop.

- Suitable speed Torque Characteristics

  Motor should have a speed torque curve similar to that of traction load. (Traction load requires high starting torque but motors like DC shunt motor, 1 ph. induction motor etc. cannot give high starting torque. This means that the speed-torque curve of DC shunt motor, 1 ph induction motor does not match with speed – torque curve of load. Hence, we cannot use this motors for traction work).

- Parallel Running

  More than one motor are required for traction purpose so, we require such motors which can easily operate in parallel.

- Overload Capacity

  Motor selected for traction purpose should have sufficient overload capacity.

- Size of Motor

  Motor selected for traction purpose should have small size with high power.

- High Efficiency

  Motor selected for traction purpose should have high mechanical and high electrical efficiency

- Braking

  Regenerative braking has many advantages. So, motor should be such that Regenerative braking can be easily used in it.

- Small Disturbance in supply

  Traction motor should handle small disturbance in supply.

Suitability of Various Motors for Traction

D.C. Series Motor

- DC series motor has high starting torque. This is one of the most required characteristics of DC series motor.
- Speed control of DC series motor is easy and various methods are available for this.
- DC series motor has speed – Torque curve similar to that of traction load.
- Size of DC series motor is small. (I.e. small size and high power rating).
- This motor can handle overload easily.
- Regenerative braking can be applied to DC series motor. But with some modifications.
- Maintenance cost required is less.
- This motor is unaffected by change in supply voltage. (I.e. voltage fluctuations)DC series motor cannot fulfill some requirements.
- Parallel running of 2 DC series motor is difficult.
- Small disturbance in supply causes Inrush current (current with 2nd harmonics and high in magnitude)Hence, D.C. series motors are better for urban and suburban services.

Three Phase Induction Motor

- Simple and rigid (strong) construction.
- Less maintenance cost.
- High efficiency.
- Good parallel operation.
- Regenerative Braking is easily done for 3 ph. Induction motor.

Phase Induction Motor Cannot Fulfill some Requirements

- High starting current is drawn by Induction motor. When motor is started at low frequency, it draws less starting current. Hence, this disadvantage is overcome.
- Speed control of Induction motor was very difficult till 1980. But with development of electronic devices like SCR, inverters we can easily change speed of Induction motor. Hence, this disadvantage is overcome.
- 3 phase induction motor requires 3 phase supply. But transmission of 3 phase supply requires 3 wires or 2 wires (rail is used as 3rd wire). So it is costly.But in kando system, transmission is done by 1 phase system and phase converter drives 3 phase induction motor. Hence, this disadvantage of costly transmission is overcome by kando system. Hence, 3 phase Induction motors are suitable for sub-urban railway system.

Traction Motor Control

Following operations are performed while controlling traction motor:

- Reduce starting current (soft starting of motor)
- Smooth acceleration to avoid sudden shock so that there will be minimum damage to the couplings.
- Providing the variable speed according to route (track).

Following methods are used for traction purpose to get above 3 requirements:

- Series parallel control of DC series motor
- For traction purpose, two motors are operated in following steps.

Step 1

- Two traction motors M1 and M2 are started with all starting resistances in series.

Step 2

- The starting resistances are cut out one by one and finally two motors are in series without any resistance.

- In series connection the supply voltage V is divided in two motors. (Both motors get half or (V/2) volts). So speed is also half.

Step 3

Motors are now connected in parallel with series resistance otherwise motor will draw very high current and may damage itself.

Step 4

Both motors are now connected in complete parallel and starting resistances are cut out one by one.

In parallel connection, voltage across M1 and M2 will be full i.e. V (voltage is always same in parallel).

As, speed × voltage

So, voltage is now increased from (V/2) to V.

Hence, speed also increases and motor runs with full speed.

Limitations

Due to the maximum recharging rate of the circuit and the capacity of battery, the braking force from an electromagnetic type RBS is always limited. Therefore, a traditional friction brake system is required to convert the excess energy from the vehicle. The friction brake can also prevent the loss of braking ability in the case of RBS failure.

RBS can only be installed on driving wheels since a drive train is required for energy recovery. The waste heat is not significantly reduced unless the vehicle is an all wheel drive model.

Adding a RBS to a vehicle means to increase the curb weight of it. Although RBS can improve fuel economy under start-and-stop driving conditions, it may have negative effect on fuel consumption during highway cruising.

The design of RBS involves varieties of sensors and logic control units to adjust the operation of RBS. The reliability concern of these electrical parts should not be neglected.

Electric Traction Systems

The electric traction system is the most efficient of all other traction system such as steam and internal combustion (IC) engine type systems. It offers several benefits over other systems, including quick start and stop, very efficient, pollution-free, easy to handle and easy speed control.

With the advancement of electrical drives for traction systems, this electrification of traction becomes much popular in several traction services including metro or suburban railways.

A system which causes the propulsion of vehicle in which tractive or driving force is obtained from various devices such as diesel engine drives, steam engine drives, electric motors, etc. is called as traction system.

It can also be defined as the railway vehicle that provides the necessary traction power to move the train is referred as the traction or locomotive. This traction power can be diesel, steam or electric power.

The traction system can be classified as non-electric and electric traction systems.

Non-electric Traction System

A traction system that doesn't use electrical energy for the movement of vehicle at any stage is referred as non-electric traction system.

The steam engine drive is the best example of a non electric traction system and it is the first locomotive system used before the invention of actual electric traction systems.

Non-electric Traction System

The steam locomotive system uses the superheated steam to produce mechanical energy for the movement of vehicle.

This may use coal or petroleum as fuel, liberates thermal energy to produce the steam pressure and then it is converted into kinetic energy so that mechanical movement of the vehicle is produced.

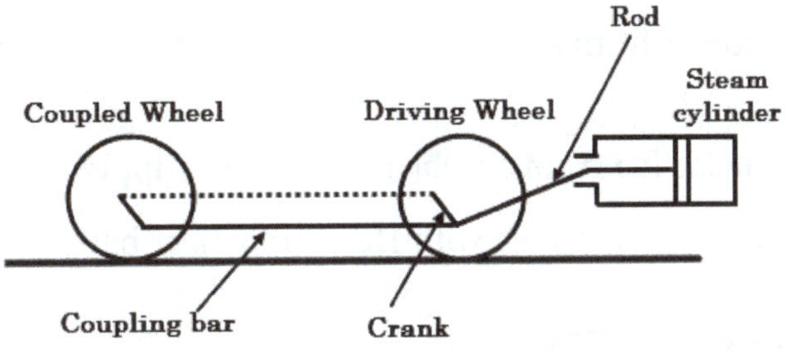

The disadvantages of steam locomotive systems, such as, low fuel efficiency, poor technical performance, maintenance of a large number of water supply facilities, and high maintenance cost makes them to be replaced by alternative traction systems and hence the electric traction.

The following are the two types of non electric traction systems:

- Steam engine drive based vehicles (used for railways)

- Internal combustion (IC) engine drive based vehicles (used for road transport)

Electric Traction System

Electric traction involves the use of electricity at some stage or all the stages of locomotive movement. This system includes straight electrical drive, diesel electric drive and battery operated electric drive vehicles.

In this, electrical motors are used for producing the vehicle movement and are powered by drawing electricity from utilities or diesel generators or batteries.

Electric Traction System

It has many advantages over non-electric traction systems such as more clean and easy to handle, no need of coal and water, easy speed control, high efficiency, low maintenance and running costs, etc.

As mentioned above, electric traction systems can be self contained locomotives or vehicles that receive power from electric distribution system (substations). Self contained locomotives includes:

- Battery operated electrical drives
- Diesel operated electrical drives

Vehicles that receives the power from substation is also referred as a third rail systems which includes:

- Railway electric vehicles fed from overhead AC or DC supply
- Trolley buses or tramways supplied with DC supply (i.e., battery electric drives)

Supply Systems of Electric Traction

The way of giving the power supply to locomotive unit is generally referred as traction electrification system. Presently, there are four types of track electrification systems are available based on the availability of supply. These are,

- DC traction system
- Single phase AC traction system

- Three phase AC traction system
- Composite traction system

DC Traction System

In this traction system, electrical motors are operates on DC supply to produce necessary movement of the vehicle. Mostly DC series motors are used in this system. For trolley buses and tramways, DC compound motors are used where regenerative braking is required.

The various operating voltages of DC traction system include 600V, 750 V, 1500V and 3000V:

- DC supply at 600-750V is universally employed for tramways and light metros in urban areas and for many suburban areas. This supply is obtained from a third rail or conductor rail, which involves very large currents.
- DC supply at 1500- 3000 is used for main line services such as light and heavy metros. This supply is drawn mostly from an overhead line system that involves small currents.

In both cases, only one conductor or rail is required to supply power to locomotive while track rails are used as return conductors in majority of cases.

Both these supply voltages are fed from substations which are located 3-5 KM for suburban services and 40 to 50KMs for main line services. These substations receive power (typically, 110/132 KV, 3 phase) from electric power grids.

This three phase high voltage is stepped-down and converted into single phase low voltage using scott-connected three phase transformers.

This single phase low voltage is then converted into DC voltage using suitable converters or rectifier such as power electronic converter, rotary converters, mercury arc converters, etc. The DC supply is then applied to the DC motor via suitable contact system and additional circuitry.

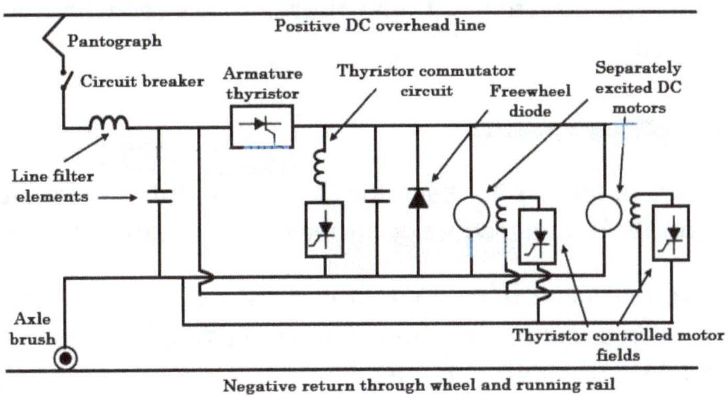

The advantages of this system include:

- In case of heavy trains that require frequent and rapid accelerations, DC traction motors are better choice as compared AC motors.
- DC train consumes less energy compared to AC unit for operating same service conditions.

- The equipment in DC traction system is less costly, lighter and more efficient than AC traction system.
- It causes no electrical interference with nearby communication lines.

Despite all these advantages, DC electrical system necessitates AC to DC conversion substations relatively at very short distances. This is the main disadvantage of DC traction system.

That's why this system is preferred only for suburban and road transport services wherein stops are frequent and also distance between stops is small.

Single Phase AC Traction System

In this type of traction system, AC series motors are used to produce the propulsion of vehicle. This system uses AC voltages from 15-25KV at a frequency of 16.7 (i.e., 16 2/3) or 25 Hz. This low frequency leads to give better performance and more efficient operation by the series motor.

This single phase supply is fed to the locomotive unit via a single overhead line while track provides the return path.

The high voltages (15-25KV) obtained from overhead conductor are stepped down to a suitable motor operating range (typically 300-400V range) using step-down transformer carried by the locomotive unit itself.

The secondary tapping of this transformer offer variable voltage to AC series motor for speed regulation.

The low frequency operation of overhead line reduces the communication interferences. Also, the reactance of the line is low at lower frequency and hence the voltage drop in the line is reduced.

Because of this low line voltage drop, it is feasible to locate the substations at 50 to 80kms apart from each other. Therefore, this system is preferred for main line services where cost of overhead system is not a much important factor and for suburban services where rapid acceleration and retardation are not required.

Three Phase AC Traction System

In this, three phase induction motors are used for the movement of locomotive. This system normally works on 3000-3600V AC at a frequency either 16 2/3 or normal supply frequency.

This system employs two overhead lines for two phases, whereas the track forms third phase. These conductors are powered from substations which are rated at higher voltages and they receive power from three-phase transmission lines.

The high voltages from transmission lines are stepped down to 3.3 KV (3000-3600 V) by transformers while the frequency is reduced by frequency converters installed at substations.

The three phase induction motor used in this system has the following characteristics; simple and robust construction, provision of regenerative braking without additional equipment and high operating efficiency, better performance, etc.

However, these motors are suffer with some drawbacks such as high starting current, low starting torque, complicated overhead structure, especially at crossings and junctions and not suitable constant speed characteristics of induction motor traction work.

These systems are adopted where high output power is required and also where automatic regeneration braking is needed. However, these systems do not found much favour compared to other systems.

Composite Traction System

The above discussed methods have their own merits and demerits. Single phase AC system has less distribution cost whereas DC system has excellent driving capability by DC series motors and three phase system has automatic regenerative braking capacity.

So by combining the advantages of AC/DC and single/three phase systems, the overall performance of the traction system gives better result than individual system and hence the evolution of composite system.

Basically composite systems are of two types, namely:

- Single phase to three phase system
- Single phase to DC system

Single Phase to Three Phase System

This traction system is also called Kando system. It consists of single phase16KV, 50 Hz supply which is fed from the substation and is being carried through a single overhead conductor.

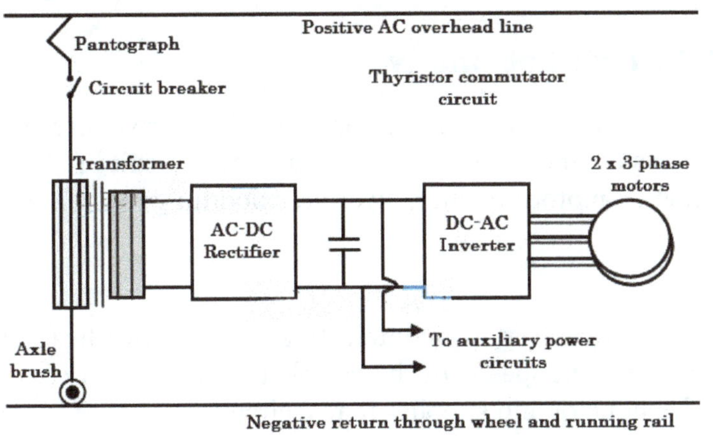

The single phase supply is then converted into three phase supply of the same frequency using phase converter equipment in the locomotive itself. The three phase supply is then fed to induction motors to drive the locomotive.

It is also possible to develop high starting torque of induction motors by reducing the supply frequency at ½ to 9 Hz by means of inverter controlled through silicon controlled rectifiers.

The main advantage of this system is that the overhead two conductor arrangement of three phase AC system is reduced to a single overhead conductor and hence more economical.

Single Phase to DC System

This traction system is most popular and widely used system everywhere. It combines the single phase high voltage AC distribution at industrial frequency with DC series motor traction.

In this, the overhead line carries single phase, 25KV, 50 Hz supply which is then stepped down to a desired range using step-down transformer located in the locomotive unit itself.

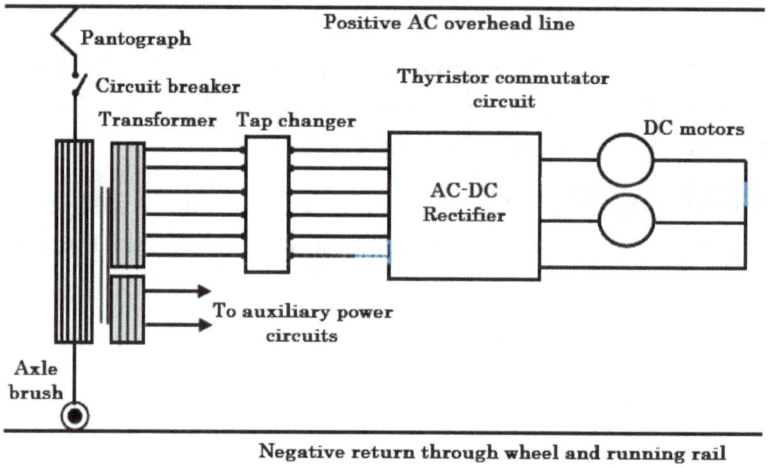

This single phase supply is then converted into DC using rectifier (in the locomotive) and then applied to DC series motor.

The advantages of this system include higher starting efficiency, less number of substations, simple substation design and lower cost of fixed installations.

Main Parts of AC Electric Locomotive

The AC locomotive system that employs single phase supply to drive three phase motor. The various components of this system include overhead contact wire, circuit breakers, pantograph, transformer, three phase traction motor, rectifier, inverter, smoothing reactor, etc.

Pantograph

The main function of pantograph is to maintain link between overhead conductor and power circuit of locomotive at different speeds of the vehicle under all wind conditions. It collects the current from overhead conductor and supplies to rest circuit.

Circuit Breaker

It protects the power circuit in the event of any fault by isolating it from the supply. It also isolates the circuit during maintenance.

Transformer

It receives the high voltage from overhead conductor via pantograph and circuit breaker and then step-down the voltage to desired level required by the rest circuit.

Rectifier

It converts a low voltage AC supply from the secondary of transformer to a DC supply.

DC Link

It connects the rectifier and inverter circuits. It consists of filter arrangement (capacitor and inductor arrangement) that filters the output from rectifier (by removing the harmonics form it) and then supplies it to the inverter.

Main Inverter

It converts the DC power to three phase AC power in order to drive three phase AC motors.

Axle Brush

It acts as a return path for the supply. Once the power is drawn to the locomotive from overhead system, the current complete its path through axle brush and one of running tacks.

Auxiliary Inverter

This inverter supplies the power to other parts in the locomotive unit including fans, motor blowers, compressors, etc.

Battery

It supplies the necessary starting current and also power up the essential circuits such as emergency lighting.

Compressor

It maintains the cooling/heating requirement in the locomotive unit.

Cooling Fans

These fans maintain the necessary cooling for the power circuits. Modern locomotive systems use electronically controlled air management systems to keep the desired temperature.

Electric Motor

Electric Motor and their Models

Presently, brushed DC motor, brushless DC motor, AC induction motor, permanent magnet synchronous motor (PMSM) and switched reluctance motor (SRM) are the main types of motors used for electric vehicle driving. The selection of motor for a specific electric vehicle is dependent on many factors, such as the intention of the EV, ease of control, etc.

In control of electric vehicle, the control objective is the torque of the driving machine. The throttle position and the break is the input to the control system. The control system is required to be fast reponsive and low-ripple. EV requires that the driving electric machine has a wide range of speed regulation. In order to guarantee the speed-up time, the electric machine is required to have large torque output under low speed and high over-load capability. And in order to operate at high speed, the driving motor is required to have certain power output at high-speed operation.

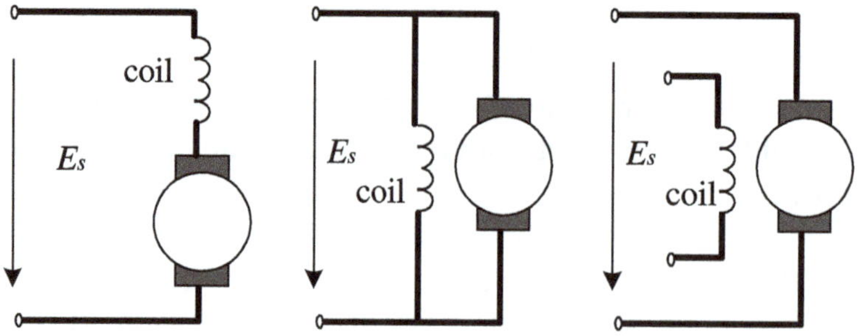

Fig: Three types of winding configuration in DC motor

Brushed DC motor

Due to the simplicity of DC motor controlling and the fact that the power supply from the battery is DC power in nature, DC motor is popularly selected for the traction of electric vehicles. There are three classical types of brushed DC motor with field windings, series, shunt and 'separately excited' windings. The shunt wound motor is particularly difficult to control, as reducing the supply voltage also results in a weakened magnetic field, thus reducing the back EMF, and tending to increase the speed. A reduction in supply voltage may, in some circumstances, have very little effect on the speed. The separately excited motor allows one to have independent control of both the magnetic flux and the supply voltage, which allows the required torque at any required angular speed to be set with great flexibility. A series wound DC motor is easy to use and with added benefit of providing comparatively larger startup torque. A (series) DC motor can generally be modeled as:

$$\begin{cases} \left(L_a + L_{field}\right)\dfrac{di}{dt} = V - \left(R_a + R_f\right)i - L_{af}i.\omega \\ J\dfrac{di}{dt} = L_{af}i^2 - B\omega - T_L \end{cases}$$

Where, i is the armature current (also field current); is the motor angular speed; $L_{a,}$ $R_{a,}$ $L_{field,}$ R_f are the armature inductance, armature resistance, field winding inductance and field winding resistance respectively; V is the input voltage, as the control input; L_{af} is the mutual inductance between the armature winding and the field winding, generally non-linear due to saturation; J is the inertia of the motor, including the gearing system and the tyres; B is the viscous coefficient; and T_L is representing the external torque, which is quantitatively the same as the one aforementioned.

Brushless DC Motor

The disadvantages of brushed DC motor are its frequent maintenance and low life-span for high intensity uses. Therefore, brushless DC (BLDC) motor is developed. Brushless DC motors use a rotating permanent magnet in the rotor, and stationary electrical magnets on the motor housing. A motor controller converts DC to AC. This design is simpler than that of brushed motors because it eliminates the complication of transferring power from outside the motor to the spinning rotor. Brushless motors are advantageous over brushed ones due to their long life span, little or no maintenance, high efficiency, and good performance of timing. The disadvantages are high initial cost, and more complicated motor speed controllers.

A BLDC motor is composed of the motor, controller and position sensor. In the BLDC motor, the electromagnets do not move; instead, the permanent magnets rotate and the armature remains static. The rotor magnetic steel is radially placed, and the permanent magnets (generally Neodymium-iron-boron: NdFeB) are installed on the surface. The magnetic permeability of such permanent magnets is close to that of air, hence can be regarded as part of the air gap. Hence, there is no salient pole effect, so that the magnetic field across the air gap is uniformly distributed. The position sensor functions like the commutator of brushed DC motor, reflecting the position of the rotor and determining the phase of current and space distribution of magnetic force.

The BLDC motor is actually an AC motor. The wires from the windings are electrically connected to each other either in delta configuration or wye ("Y"-shaped) configuration. In figure, an equivalent circuit of wye-connected BLDC is shown. With this configuration, the simplified model can be obtained as:

$$\begin{bmatrix} u_a \\ u_b \\ u_c \end{bmatrix} = \begin{bmatrix} R & 0 & 0 \\ 0 & R & 0 \\ 0 & 0 & R \end{bmatrix} \begin{bmatrix} i_a \\ i_b \\ i_c \end{bmatrix} + \begin{bmatrix} L & 0 & 0 \\ 0 & L & 0 \\ 0 & 0 & L \end{bmatrix} . P \begin{bmatrix} i_a \\ i_b \\ i_c \end{bmatrix} + \begin{bmatrix} e_a \\ e_b \\ e_c \end{bmatrix} + \begin{bmatrix} U_N \\ U_N \\ U_N \end{bmatrix}$$

Where, $L = L_S - M; L_S$: self-inductance of the windings; M : mutual inductance between two windings; R:

stator resistance per phase; u_a, u_b, u_c: stator phase voltages; i_a, i_b, i_c: stator phase currents; e_a, e_b, e_c the back emfs in each phase; $P = \frac{d}{dt}$.

The generated electromagnetic torque is given by:

$$T_e = \frac{P_e}{\omega} = (e_a i_a + e_b i_b + e_c i_c)/\omega$$

And the kinetics of the motor can be described as:

$$T_e - T_L - f\omega = j\frac{d\omega}{dt}$$

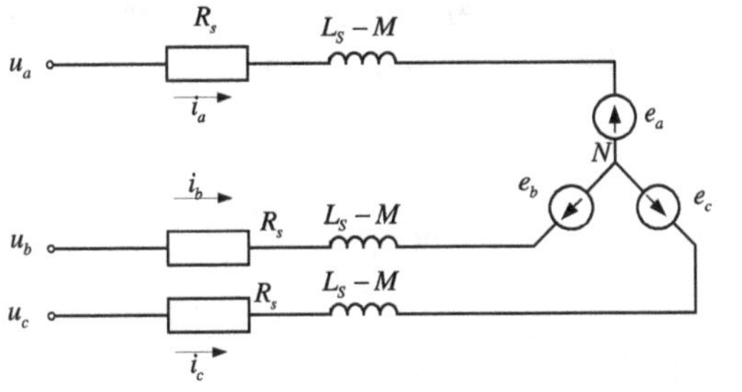

Fig: Equivalent circuit of BLDC

Where, ω: the angular velocity of the motor; $T_{e,} T_L$: electromagnetic torque of the motor and the load torque;

P_e : electromagnetic power of the motor; J: moment of inertia; f: friction coefficient.

Under normal operation, only two phases are in conduction. Then the voltage balance equation, back EMF equation, torque equation, and kinetic equations that govern the operation of a WYE connected BLDC motor can be obtained as:

$$\begin{cases} u_d = E + i \cdot R + L \cdot \dfrac{di}{dt} \\ E = K_e \cdot n \\ T_e = K_T \cdot n \\ T_e = T_L + f\omega + J\dfrac{d\omega}{dt} \end{cases}$$

Where, u_d: the voltage across the two windings under conduction; E: the back EMF of the two windings under conduction; K_T: torque coefficient, and K_e: back EMF coefficient. It is shown that, in BLDC motors, current to torque and voltage to rpm are linear relationships.

Permanent Magnet Synchronous Motor

A permanent magnet synchronous motor is a motor that uses permanent magnets to produce the air gap magnetic field rather than using electromagnets. Such motors have significant advantages, such as high efficiency, small volume, light weight, high reliability and maintenance-free, etc., attracting the interest of EV industry. The PMSM has a sinusoidal back emf and requires sinusoidal stator currents to produce constant torque while the BDCM has a trapezoidal back emf and requires rectangular stator currents to produce constant torque. The PMSM is very similar to the wound rotor synchronous machine except that the PMSM that is used for servo applications tends not to have any damper windings and excitation is provided by a permanent magnet instead of a field winding. Hence the d, q model of the PMSM can be derived from the well known model of the synchronous machine with the equations of the damper windings and field current dynamics removed.

In a PMSM, the magnets are mounted on the surface of the motor core. They have the same role as the field winding in a synchronous machine except their magnetic field is constant and there is no control on it. The stator carries a three-phase winding, which produces a near sinusoidal distribution of magneto motive force based on the value of the stator current. In modeling of rotating machines like PMSM, it is a general practice to perform Park transform and deal the quantities under dq framework. The dqo transform applied to a three-phase quantities has following form:

$$\begin{bmatrix} x_d \\ x_q \\ x_o \end{bmatrix} = \sqrt{\frac{2}{3}} \cdot \begin{bmatrix} \cos\theta & \cos\left(\theta - \dfrac{2\pi}{3}\right) & \cos\left(\theta + \dfrac{2\pi}{3}\right) \\ -\sin\theta & -\sin\left(\theta - \dfrac{2\pi}{3}\right) & -\sin\left(\theta + \dfrac{2\pi}{3}\right) \\ \sqrt{\dfrac{1}{2}} & \sqrt{\dfrac{1}{2}} & \sqrt{\dfrac{1}{2}} \end{bmatrix} \cdot \begin{bmatrix} x_a \\ x_b \\ x_c \end{bmatrix}$$

Where, the x can be voltage u or current *i*. Under the dqo framework, the equivalent circuit of d-axis and q-axis circuits of a PMSM motor is shown in figure, and the model of a PMSM can be written as:

$$\begin{cases} u_d = r_s i_d + p\Psi_d - \omega\Psi_q \\ u_q = r_s i_d + p\Psi_d - \omega\Psi_q \\ \Psi_d = L_d i_d + \Psi \\ \Psi_q = L_q * i_q \\ T_e = 1.5_m\left[\Psi i_q + \left(L_d - L_q\right) i_d i_q\right] \\ J\dfrac{dw_r}{dt} = T_e - B\omega_r - T_L \end{cases}$$

Where, $\Psi_{d,}\Psi_q$: the flux linkages of d-axis and q-axis respectively; L_d, L_q: self inductance of dq axes; i_d, i_q: d_q-axis current; u_d, u_q: d_q-axis voltage; ω_r: angular velocity of rotor; r_s: stator resistance; p_m: number of poles; Ψ: flux linkage produced by the rotor permanent magnet; T_e: motor torque; T_L: load torque; *J*: moment of inertia; *B*: friction coefficient; p: differential operator. The first two equations are the equations for stator voltage, the next two equations are about the magnetic flux linkage, the fifth equation is about the calculation of torque, and the last equation is about the kinetics of the motor.

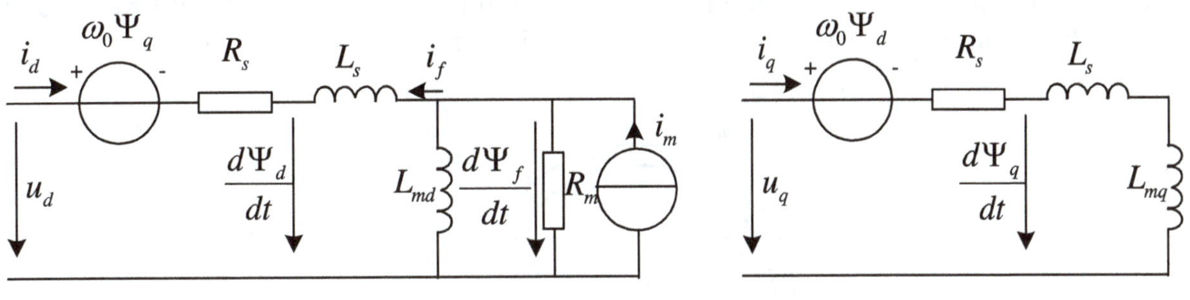

Fig: d-axis and q-axis equivalent circuit model of PMSM

Induction Motor

Induction machines are among the top candidates for driving electric vehicles and they are widely used in modern electric vehicles. Some research even concludes that the induction machine provides better overall performance compared to the other machines.

An induction motor (or asynchronous motor or squirrel-cage motor) is a type of alternating current motor where power is supplied to the rotor by means of electromagnetic induction. It has the advantages such as low-cost, high-efficiency, high reliability, maintenance-free, easy for cooling and firm structure, etc. making it specially competitive in EV driving. In induction motor, stator windings are arranged around the rotor so that when energised with a polyphase supply they create a rotating magnetic field pattern which sweeps past the rotor. This changing magnetic field pattern induces current in the rotor conductors, which interact with the rotating magnetic field created by the stator and in effect causes a rotational motion on the rotor.

AC induction motor is a time-varying multi-variable nonlinear system, hence the modeling task is not easy. For simplicity, following assumptings have to be made:

- Magnetic circuit is linear, and saturation effect is neglected;
- Symmetrical two-pole and three phases windings (120° difference) with edge effect neglected;
- Slotting effects are neglected, and the flux density is radial in the air gap and distributed along the circumference sinusoidally;
- Iron losses are neglected.

With such assumptions, the physical model of an induction motor can be given as shown in figure below. The three-phase stators are fixed on A, B and C axes, which are stationary reference frames. The three-phase rotor windings are fixed on a, b and c axes, which are rotating frames. Hence the equations governing the dynamics of the induction motor can be given as:

$$u = R_i + L\frac{di}{dt} + \omega_r \frac{\partial L}{\partial \theta_0} i$$

Where, $u = [u_A, u_B, u_C, u_a, u_b, u_c]^T$, vector of stator and rotor voltages; $i = [i_A, i_B, i_C, i_a, i_b, i_c]^T$ vector of stator and rotor current; $\omega_r = d\theta_0/dt$: the angular speed of rotation; *J*: the total moment of inertia; T_L : load torque; $R = diag[\,R_1\ R_1\ R_1\ R_2\ R_2\ R_2\,]$: where, R_1 is the resistance of stator winding and R_2 is the resistance of rotor winding; and rotor; *LAB*, *Lab*: mutual inductance of stators and rotors respectively; *M*: mutual inductance between stator and rotor.

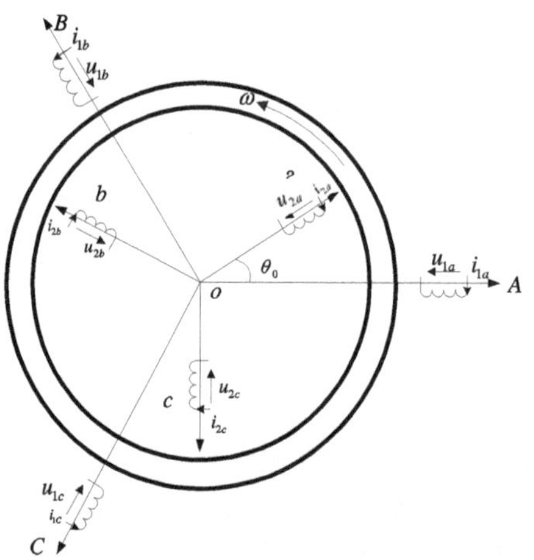

Fig: Physical Model of 3-phase AC induction Motor

$$L = \begin{bmatrix} L_{11} & L_{12} \\ L_{21} & L_{22} \end{bmatrix}, \text{ where } L_{11} = \begin{bmatrix} L_A & L_{AB} & L_{AB} \\ L_{AB} & L_A & L_{AB} \\ L_{AB} & L_{AB} & L_A \end{bmatrix}, \quad L_{22} = \begin{bmatrix} L_a & L_{ab} & L_{ab} \\ L_{ab} & L_a & L_{ab} \\ L_{ab} & L_{ab} & L_a \end{bmatrix} \text{ and}$$

$$L_{12} = L_{12}{}^T = M\begin{bmatrix} \cos\theta & \cos(\theta - 120^\circ) & \cos(\theta + 120^\circ) \\ \cos(\theta - 120^\circ) & \cos\theta & \cos(\theta - 120^\circ) \\ \cos(\theta + 120^\circ) & \cos(\theta - 120^\circ) & \cos\theta \end{bmatrix}; L_{A,}\ L_a : \text{self-inductance of stator}$$

The first equation is the voltage equation and the second equation is the kinetic equation of the motor.

Controller design of electric vehicle driven by different motors.

Figure shows a universal framework for electric vehicle controller. The vehicle is driven by a motor, which is supplied by the battery through a controlled power circuit. Other than circuit for control of the motor, there is quite a lot of auxiliary control for auto electronics. The control strategies are implemented in the microprocessor, such as DSP (Digital Signal Processor).

Control of electric vehicle is essentially the control of motor. In figure, only the motor and its associated driving power circuit will be replaced with different motors. With different motors, it is necessary to use different control strategies. However, it is not possible to include all type of motor and control strategies in one book. Hence, in this chapter, only one typical controller or control strategy will be presented. It is noticed that generally PWM control is used for DC motor, while variable-voltage variable-frequency (VVVF), FOC (field-oriented control) and DTC (direct torque control) are used for induction motor. And some traditional control algorithms, such as PID, cannot satisfy the requirements of EV control. Many modern high-performance control technologies, such as adaptive control, fuzzy control, artificial neuro network and expert system are being used in EV controllers.

Driven by Brushed DC Motor

The controller design for an EV driven by series wound DC motor will be discussed. When the electric vehicle is driven by a series wound DC motor:

$$\begin{cases} \left(L_a + L_{field}\right)\dfrac{di}{dt} = V - \left(R_a + R_f\right)i - L_{af}i.\omega \\ \left(J + m\dfrac{r^2}{G^2}\right)\dfrac{d\omega}{dt} = L_{af}i^2 - B\omega - \dfrac{r}{G}\left(\mu_{rr}mg + \dfrac{1}{2}\rho AC_d\upsilon^2 + mg\sin\phi\right) \end{cases}$$

In this case, a model-based controller can be designed. Unlike other applications in which the system generally operates around the equilibrium point, the operation of EV may take a very wide range (e.g., from zero to full speed). Hence, it is essentially to design EV controller with nonlinear control techniques. The model-based controller is very sensitive to the uncertainties in the parameters. Many parameters in the complex vehicle dynamics cannot be precisely modeled and some parameters may vary due to the varying operation conditions.

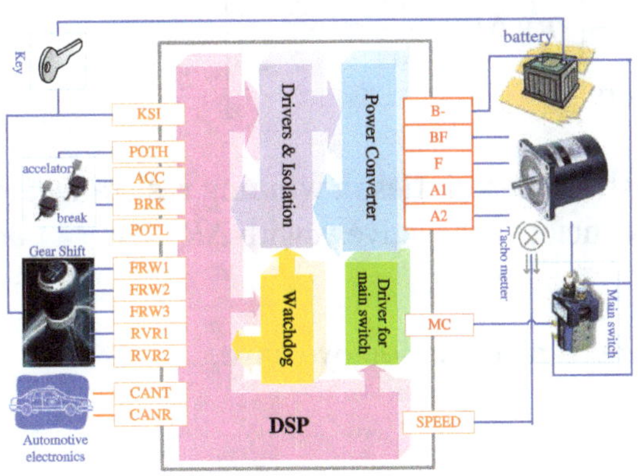

Model of electric vehicle controller

For example, the resistance in the armature winding of a motor would change as the operation temperature varies. Hence, when designing the controller, the robustness of the controller should be first considered.

For the convenience of designing nonlinear controller, first change the model in into the following format:

Where,

$$X=\begin{bmatrix}x_1\\x_2\end{bmatrix}=\begin{bmatrix}i\\\omega\end{bmatrix};f(X)=\begin{bmatrix}-\frac{R_a+R_f}{L_a+L_{field}}x_1-\frac{L_{af}}{L_a+L_{field}}x_1.x_2\\\frac{1}{J+m\frac{r^2}{G^2}}\left\{L_{af}x_1^{\ 2}-Bx_2-\frac{r}{G}\left(\mu_{rr}mg+\frac{1}{2}\rho AC_d\frac{r^2}{G^2}x_2^{\ 2}+mg\sin\phi\right)\right\}\end{bmatrix};$$

$$g(X)=\begin{bmatrix}\frac{1}{L_a+L_{field}}\\0\end{bmatrix};h(X)=x_2.$$

In order to consider the uncertainties of the system:

$$\begin{cases}X=f(X)+q(X,\theta(t))+g(X)u\\y=h(X)\end{cases}$$

$f(X)$, $g(X)$ and $h(X)$ are the same. $q(X,\theta(t))$ is used to include the model uncertainties, where $\theta(t)$ is the uncertainty vector. In the model described above, the rolling resistance coefficient and aerodynamic dragging coefficients cannot be precisely modeled. These coefficients are always varying along the moving of the vehicle (e.g., due to wind). Also, the resistance of the windings is also varying due to the variation of temperature. Hence, $q(X,\theta(t))$ can be modeled as:

$$q(X,\theta(t))=\begin{bmatrix}\frac{i}{L_a+L_{field}}\Delta R\\-\frac{r/G}{J+m\frac{r^2}{G^2}}\left(mg\Delta\mu_{rr}+\frac{r^2}{G^2}\omega^2\Delta C_{ad}\right)\end{bmatrix}$$

Where, Δ_R, $\Delta\mu_{rr}$ and ΔC_{ad} are the uncertainties in winding resistance, rolling resistance coefficient and aerodynamic dragging coefficient respectively, with ΔR_m, $\Delta\mu_{rr-m}$ and ΔC_{ad-m} epresenting their maximum uncertainties.

By using the similar coordinate transformation, we have,

$$\begin{cases}\dot{z}_1=\dot{\omega}-\dot{\omega}_0=z_2\\\dot{z}_2=L_f^2h(X)+L_gL_fh(X)u+L_qL_f(X)\end{cases}$$

Where,

$$L_q h(X) = -\frac{r/G}{J + m\dfrac{r^2}{G^2}}\left(mg\Delta\mu_{rr} + \frac{r^2}{G^2}x_2^2\Delta C_{ad}\right)$$

$$L_q L_f h(X) = \frac{1}{J + m\dfrac{r^2}{G^2}}\frac{2x_1^2}{L_a + L_{field}}\Delta R - \frac{r/G\left(mg\Delta\mu_{rr} + \dfrac{r^2}{G^2}x_2^2\Delta C_{ad}\right)\left(\rho AC_d\dfrac{r^2}{G^2}x_2 - B\right)}{\left(J + m\dfrac{r^2}{G^2}\right)^2}$$

Hence, by using the recursive backstepping design method with robust control system, one can select a robust control Lyapunov function (rclf) as:

$$V(z_1, z_2) = \frac{1}{2}z_1^2 + \frac{1}{2}(z_2, \tilde{z}_2)^2$$

Where, $\tilde{z}_2 = s_1 z_1$, and $s_1(z_1) = \dfrac{r^3/G^3}{J + m\dfrac{r^2}{G^2}}(z_1 + 2\omega_0)\Delta C_{ad\_m} - 2.$

Then, the control law in can robustly stabilize the electric vehicle system with any parametric uncertainties:

$$u = \frac{\tilde{z}_3 - L_f^2 h(X)}{L_g L_f h(X)}$$

Where,

$$\tilde{z}_3 = s_2(z_2 - s_2 z_1)$$

$$\begin{aligned} s_2 = \frac{T_1^2}{2} - 1 - T_1 - \beta_6\Delta R_m - \{T_1[3 + 2a_2 z_1 + 4a_2\omega_0]\Delta C_{ad\_m} \\ -\beta_6\Delta R_m s_1 - 1 + \left[\left(a_4 z_1^2 + (a_3 + 3a_4\omega_0)z_1 + (2a_3\omega_0 + 3a_4\omega_0{}^2)\right)\Delta C_{ad\_m} \\ + a_2\Delta\mu_{rr\_m} + (a_7 z_1 + a_8)\Delta R_m\right]\}^2 / 4 \end{aligned}$$

$$T_1 = (2a_1 z_1 + 2a_1\omega_0)\Delta C_{ad\_m} + 2; a_1 = (r^3/G^3)/(J + mr^2/G^2); a_2 = mg\rho AC_d(r/G)^4/(J + mr^2/G^2)^2;$$

$$a_3 = B(r/G)^3/(J + mr^2/G^2)^2; a_4 = \rho AC_d(r/G)^6/(J + mr^2/G^2)^2;$$

$$a_5 = 2L_{af}/(J + mr^2/G^2)/(L_a + L_f); a_6 = 2/(L_a + L_f); a_7 = a_5\frac{1}{2L_{af}}\rho AC_d\frac{r^3}{G^3};$$

$$a_8 = a_5\frac{1}{L_{af}}\rho AC_d\frac{r^3}{G^3}\omega_0 + a_5\frac{B}{L_{af}}.$$

Figure below shows the robustness test result of a controller designed for a light-weighted EV, compared against a regular PID controller, with parameters as specified in table. To facilitate the graphical representation, the sudden change in hill climbing angle (0.1 rad) is applied on the non-linear robust controller system at t=20 s and on the double-loop PID controller at t=50 s. Shown in figure is the excellent robust performance of the designed controller under parameter uncertainties. The test is performed with the arbitrary combinations (i.e., uncertainty in single parameter, two parameters and three parameters) of ±10% uncertainty in the aforementioned three uncertain parameters. Again, for comparison purpose, the performance of (double loop) PID controller and the nonlinear optimal controller are plotted in figures.

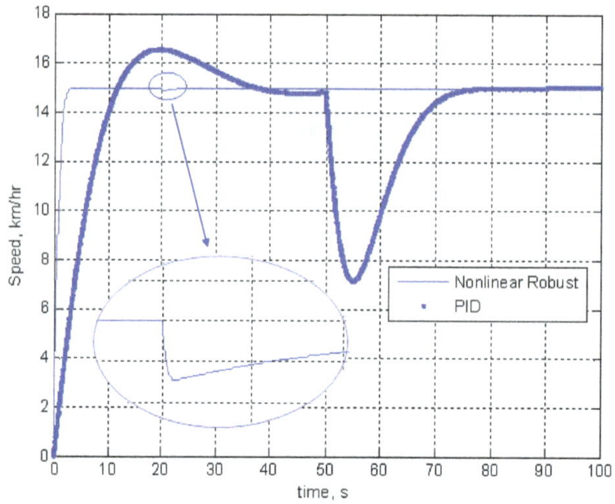

Figure: Robustness test of the non-linear robust controller

To systematically test the performance, the New European Driving Cycle (NEDC) is used. The New European Driving Cycle is a driving cycle consisting of four repeated ECE-15 driving cycles and an Extra-Urban driving cycle, or EUDC. The test results are shown in (the maximum speed is scaled to 50km/h here). It is shown that the nonlinear controller has much better tracking performance than the double loop PID controller, especially in the range of speed below designed nominal speed. And it does not increase much amp-hour consumption (nonlinear optimal: 4.48km/11.97AH; non-linear robust: 4.825km/10.78AH; PID: 4.49km/10.67AH).

	Motor		Vehicle
La + *Lf* (mH)	6.008	m (kg)	800
Ra + *Rf* (Ω)	0.12	A (m2)	1.8
B (N.M.s)	0.0002	ρ (kg/m3)	1.25
J (kg. m2)	0.05	*Cd*	0.3
Laf (mH)	1.766	φ (°)	0
V (V)	0~48	μ*rr*	0.015
i (A)	78A (250max)	*r* (m)	0.25
ωnom (r/min)	2800 (*v*=25km/h)	*G*	11

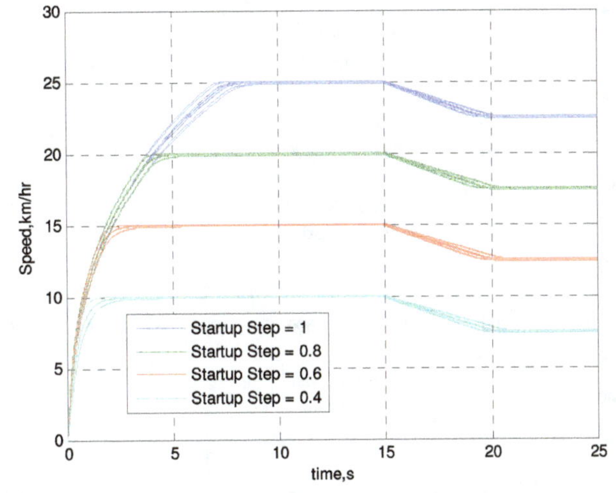

Fig:. Robustness test of the non-linear robust controller

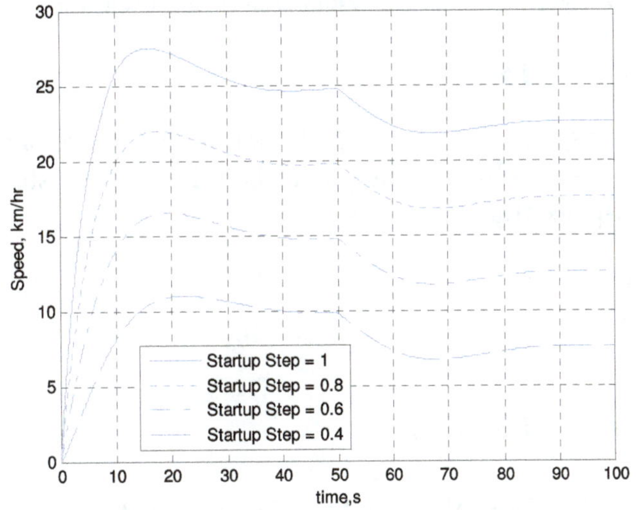

Figures: Performance of non-linear optimal controller

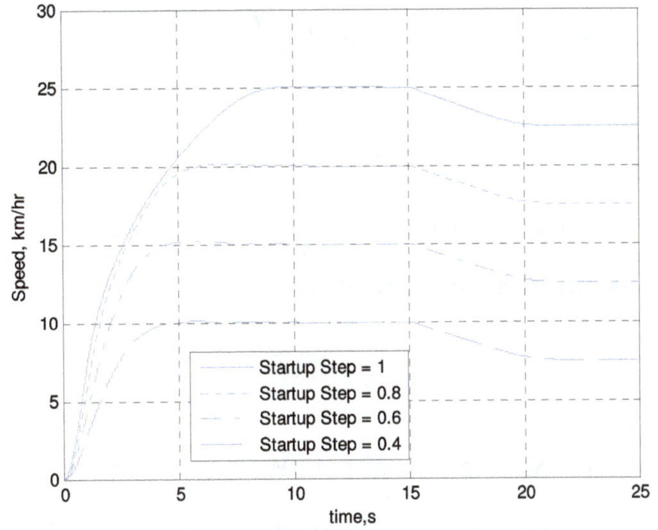

Figure: Performance of double-loop PID controller

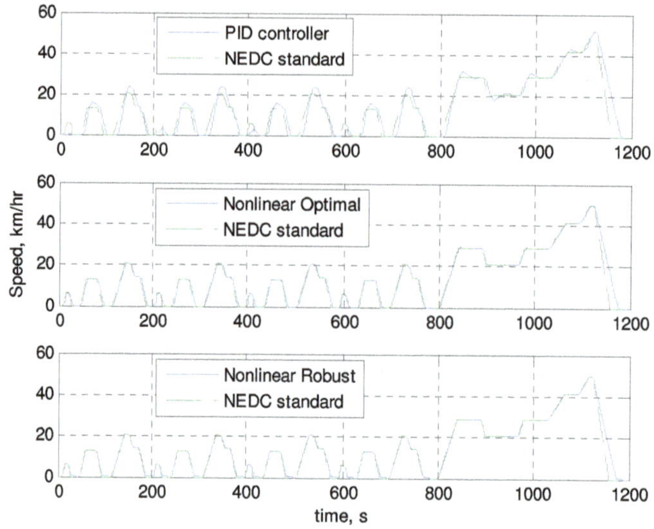

Figures: Results of new European Driving Cycle test

Driven by Brushless DC Motor

BLDC motor is a closed loop system in nature. The back EMF introduces a negative feedback signal proportional to the motor speed, which enhances the damping of the system. Assume all the initial conditions are zero, the Laplace transform of is:

$$\begin{cases} u_d(s) = E(s) + i(s).R + L.s.i(s) \\ E(s) = K_e \ .n(s) \\ T_e(s) = K_T \ .i(s) \\ T_e(s) = T_L(s) + f\omega(s) + J.s.\omega(s) \end{cases}$$

Therefore, the dynamic model of an EV driven by BLDC can be obtained as shown in figure.

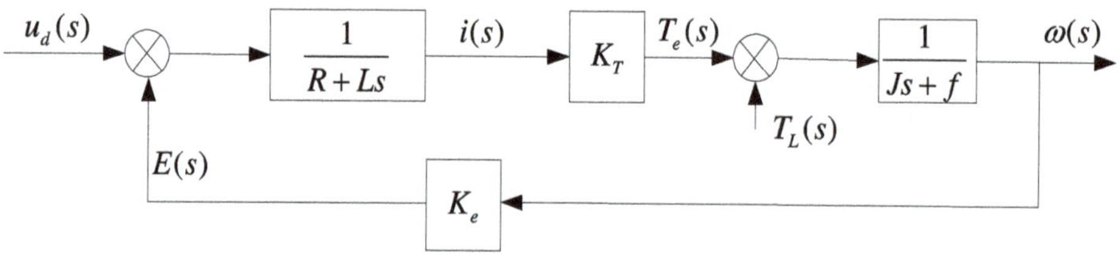

Figure: Dynamic model of an EV driven by BLDC

The transfer function of the inverter can be given as:

$$G_{pwm}(s) = \frac{K_{pwm}}{T_s.s+1}$$

Where, K_{pwm}: gain of inverter; *TS*: time constant of PWM controller.

The controller of a BLDC is generally composed of current regulation loop and speed regulation loop. In practical systems, due to the fact that the electromagnetic time constant is smaller than electromechanical time constant, current regulation is faster than speed regulation. Hence, speed regulation is faster than the variation of back EMF. Therefore, the effect of back EMF on current regulation loop can be neglected. In order to have small overshoot and good tracking performance, current regulation should be designed as type-I system. Since there are two inertia elements, the current regulator should be designed as PI regulator, whose transfer function is:

$$G_i(s) = K_{pi} \cdot \frac{T_i \cdot s + 1}{T_i \cdot s} = K_{pi} \cdot \left(1 + \frac{1}{T_i \cdot s}\right)$$

Where, K_{pi}: the proportional coefficient of current regulator; T_i: time constant of current regulator.

The structure of the current regulator is shown as the internal loop in Figue above neglecting the effect of back EMF on the current regulation loop, the stator circuit of the motor can be approximated as a first-order inertia element, hence:

$$\frac{i_s}{u_s} = \frac{1}{R + L.s} = \frac{1/R}{T_a \cdot s + 1}$$

Where, $T_a = L/R$.

The structure of the speed regulation is shown as the external loop in figure. The speed regulation system should have no steady error at steady state and good anti-disturbance capability at transient state, the speed regulator should be designed as type-II system. The speed regulation loop is composed of an integration element and an inertia element, a PI regulator should be used, leading to a transfer function:

$$G_\omega(s) = K_{P\omega} \cdot \frac{T_\omega \cdot s + 1}{T_\omega \cdot s}$$

Where, K_{pw}: proportional coefficient of speed regulator; T_w: time constant of speed regulator.

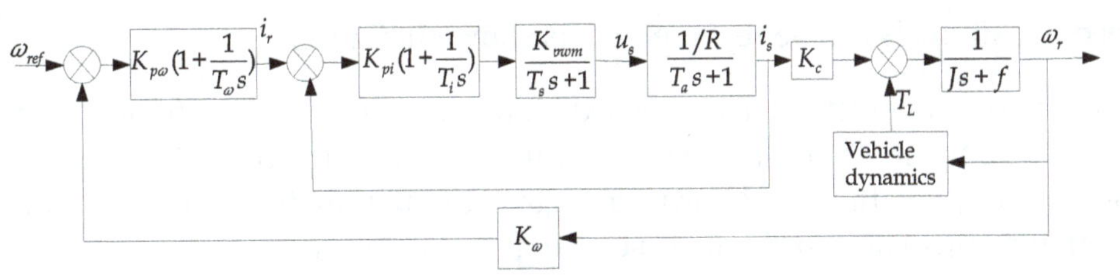

Figure: Structure of the controller for EV driven by BLDC

Driven by PMSM

When designing a controller for PMSM motor, generally two control strategies, i.e., vector control and direct torque control (DTC), are used. In DTC, Stator flux linkage is estimated by integrating the stator voltages. Torque is estimated as a cross product of estimated stator flux linkage vector and measured motor current vector. The estimated flux magnitude and torque are then compared with their reference values. If either the estimated flux or torque deviates from the reference more than allowed tolerance, the transistors of the variable frequency drive are turned off and on in such a way that the flux and torque will return in their tolerance bands as fast as possible. Thus direct torque control is one form of the hysteresis or bang-bang control. In vector control, the stator phase currents are measured and converted into a corresponding complex (space) vector. This current vector is then transformed to a coordinate system rotating with the rotor of the machine. The position can then be obtained by integrating the speed. Then the rotor flux linkage vector is estimated by multiplying the stator current vector with magnetizing inductance *Lm* and low-pass filtering the result with the rotor no-load time constant *Lr*/*Rr*, that is, the ratio of the rotor inductance to rotor resistance. Using this rotor flux linkage vector the stator current vector is further transformed into a coordinate system where the real x-axis is aligned with the rotor flux linkage vector. The real x-axis component of the stator current vector in this rotor flux oriented coordinate system can be used to control the rotor flux linkage and the imaginary y-axis component can be used to control the motor torque.

In DTC, the switching speed is low, and the controlled motor generally has low inductance. Therefore, significant current and torque ripples are observed at low speed, limiting the speed regulation range of the controlled system. While the vector control can handle such problems very well. No matter under low-speed or high-speed condition, the motor current can respond very well once the current waveform required for certain rotation speed is given. The q-axis component of the current is proportional to the torque component needed, giving an excellent performance in dynamic response. Although the vector control algorithm is more complicated than the DTC, the algorithm is not needed to be calculated as frequently as the DTC algorithm. Also the current sensors need not be the best in the market. Thus the cost of the processor and other control hardware is lower making it suitable for applications where the ultimate performance of DTC is not required. Therefore, vector control is selected in the EV control.

When the number of poles is fixed, the torque of a PMSM is determined by the stator current, therefore, the control of the motor is to control the current. Generally, the stator current *is* is first decomposed in the d-q cordinate, in which d-axis is aligned with the rotor flux and the q-axis is vertical to d-axis.

$$i_s = i_d + ji$$

After such decomposition, is can be represented as shown in figure.

It is shown that the stator current is decomposed along the rotor axis i_d component and vertical to the rotor axis i_q component. Then the control of the motor become the calculation of the instantaneous rotor position, i.e., the ε as shown in Figure, the parameter ε can be obtained through sensor. Using an intermediate coordinate system α, β and its phase

current projections i_α and i_β, and the fact that $i_a + i_b + i_c = 0$, one obtains

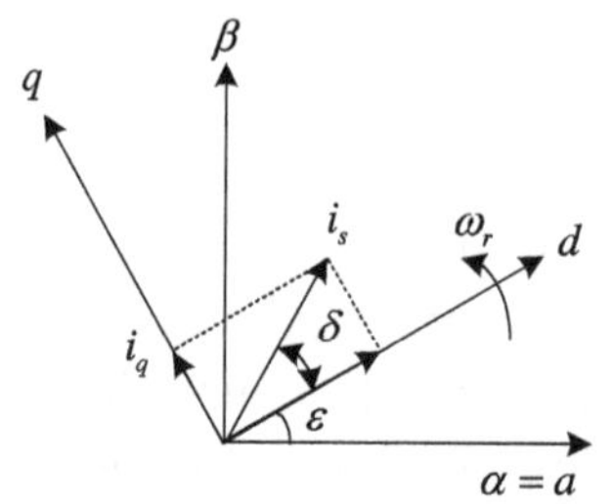

Figure: Current vector decomposition

$$\begin{cases} i_a = i_a \\ i_\beta = \frac{1}{3} i_a + \frac{2}{3} i_\beta \end{cases}$$

i_d and i_q are then deduced from i_α and i_β by a rotation of the angle ε:

$$\begin{bmatrix} i_d \\ i_q \end{bmatrix} = \begin{bmatrix} \cos \in & \sin \in \\ \sin \in & \cos \in \end{bmatrix} \begin{bmatrix} i_a \\ i_\beta \end{bmatrix}$$

With proper coordinate transformation, the electric torque of a synchronous motor is described as follows:

$$T_d = P * \Phi * i_q$$

Where, Φ : flux linkage.

It is shown that, for a magnetically isotropous machine the motor torque depends only on the quadrature *q* current component (torque component). When δ = π /2 (i.e., *id* = 0), the optimal mode of operation is achieved, where the motor produces the maximum torque.

Currently, vector control (or FOC) is an effective method in variable frequency speed regulation system in synchronous motors. It can obtain large instantaneous speed regulation range. The stator current is measured, then transformed to rotor coordinate with coordinate transformation, forming a current feedback, which can dynamically follow the variation of the current. The speed and rotor position are obtained by optical encoder, forming the speed loop, and finally implement the control of current and speed, greatly enhancing the stability, response speed and control accuracy of the system.

The key to control of PMSM is the realization of high-performance control of motor instantaneous torque. The requirements of torque control for PMSM can be summarized as fast response, high accuracy, low ripple, high efficiency and high power factor, etc. There are three regular methods in current control:

1) $i_d = 0$ control;

2) Maximum torque and current control;

3) $i_d < 0$ field weakening control.

Other methods such as constant flux linkage control or torque linear control are too complex to use.

$i_d = 0$ control is simple to implement, and the EM torque and the stator current have a linear relationship, hence having no demagnetizing effect. The disadvantages of this method is the high stator voltage and low power factor at large load conditions, hence cannot fully utilize capacity. In EV control, generally the capacity required is not very large, but it needs high overload capability and good torque response performance. Hence, *id* = 0 control, together with maximum torque-current ratio method, is selected as the solution in EV driven by PMSM. Under such conditions, the structure of the controller is designed as shown in figure.

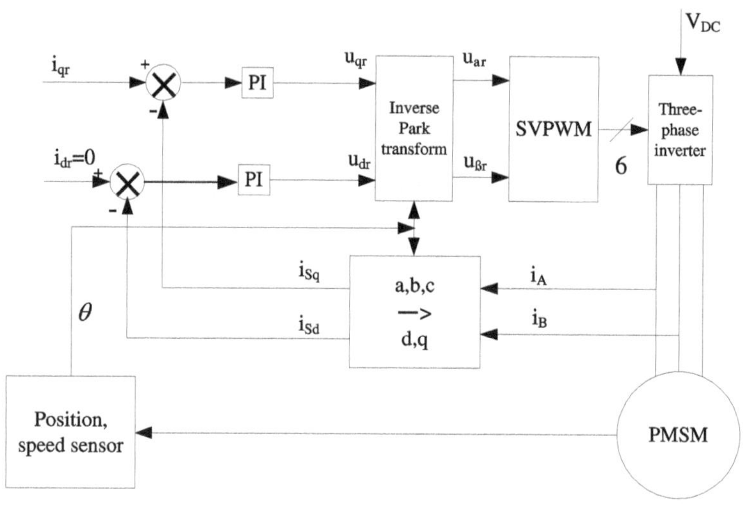

Figure: Vector control of PMSM

Optical encoder is installed the motor rotor, which accepts the pulse signal when the rotor is rotating. With the pulse signal, the space position θ and rotor speed *ωr* can be calculated.

Using the transforming factor e-jθ*r*, *id* and *iq* are obtained by performing coordinate transform from ABC-coordinate to dq-coordinate on *iA*, *iB* and *iC* (obtained by assuming *iC* = - *iA* – *iB* here). The detected *id* and *iq* are compared with the reference value *idr* and *iqr*. The error signal is used for regulation, e.g. PI regulator. The inverse Park transform is applied on the calcuated input and then generator output by looking up the table. The generated output is supplied to the three-phase inverter through SVPWM (Space Vector PWM), which controls the motor.

In the SVPWM modulation system, the gain of the inverter can be expressed as:

$$K_{VSI} = \frac{U_o}{2U_\Delta}$$

Where, U_o: DC voltage input to the inverter; U_Δ: amplitude of the triangular modulation signal.

The control object of current loop is the DC-AC inverter and the stator circuit of the PMSM motor. DC-AC inverter can generally be regarded as a first-order inertia element with time

Constant T_{PWM} ($T_{PWM} = \frac{1}{2f_\Delta}$, where f_Δ : frequency of the triangular modulation signal).

The transfer function is:

$$G_{VSI}(s) = \frac{K_{VSI}}{T_{PWM}s + 1}$$

With similar discussion in BLDC motor, the stator circuit of the motor is approximated by a first-order inertia element. Introducing a PI regulator, the system is tuned to be type-I element. The PI regulators used for d-axis and q-axis are the same, and the transfer function is:

$$PI_i = K_{Pi}\left(1+\frac{1}{T_{ii}s}\right)$$

In this case, the structure of the current regulation system is shown as the internal loop in figure. The close loop transfer function of the current regulation system can be simplified as a first-order inertia element:

$$G_{ci}(s)=\frac{G_{oi}(s)}{1+G_{oi}(s)}=\frac{K_i}{s+k_i}=\frac{1}{sk_i+1}$$

Where, $K_i=\frac{K_{pi}K_{VSI}}{T_aR_s}=\omega_c;\omega_c$: close-loop bandwidth of current loop

Hence, if a PI regulator is used for speed loop, with transfer function:

$$G_{as}(s)=K_s\left(1+\frac{1}{T_s s}\right)$$

Where, *Ks*, *Ts*: amplification coefficient and time constant of PI regulator respectively.

Then the overall model of control system is shown in figure, with the outer loop being the speed loop.

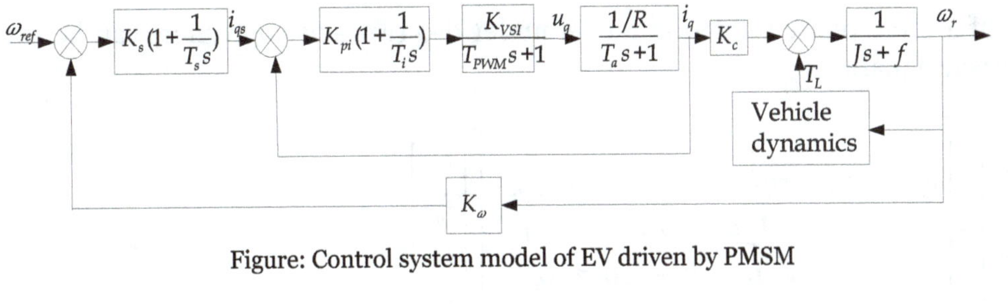

Figure: Control system model of EV driven by PMSM

Driven by Inductor Motor

Generally, VVVF based on steady motor model, DTC and vector control are used for the speed regulation for induction motor applications. The first method does not consider the complex dynamics inside the electric machine, therefore cannot achieve good dynamic performance. With DTC, although it is not affected by the motor parameters, over-current may occur due to the fact that there is no current feedback. Under low speed condition, the stator flux linkage is circular, and the current is close to sinusoidal. However, at high speed condition, the current waveform is irregular. Large harmonic currents and electromagnetic noises can be observed. Vector control, completely solved above issues. With vector control, the control of AC induction motors is as simple as that of DC motor.

The basic principle of vector control is that the magnetic force and power are invariant under normal transform. First transform the model under A-B-C coordinate to α – β steady coordinate

(Clarke transform), then apply Park transform to the model under α – β steady coordinate to d-q rotating coordinate.

Assume the windings of three phase A, B and C are symmetrical and flowing through balanced sinusoidal current. Then the resultant magnetic force F is rotating with synchronous speed ϖ_1.

Assume balanced three phase, the projection that modifies the three phase system into the (α,β) two dimension orthogonal system is:

$$\begin{cases} i_a = i_a \\ i_\beta = \frac{1}{\sqrt{3}} i_a + \frac{2}{\sqrt{3}} i_b \end{cases}$$

The (α,β)→(d,q) projection (Park transformation) is the most important transformation in the vector control. This projection modifies a two phase orthogonal system (α,β) in the d,q rotating reference frame. If we consider the d axis aligned with the rotor flux, for the current vector, the relationship from the two reference frame:

$$\begin{bmatrix} i_d \\ i_q \end{bmatrix} = C_{2s/2r} \begin{bmatrix} i_a \\ i_\beta \end{bmatrix} = \begin{bmatrix} \cos\theta & \sin\theta \\ -\sin\theta & \cos\theta \end{bmatrix} \begin{bmatrix} i_a \\ i_\beta \end{bmatrix}$$

Whose inverse transformation is:

$$\begin{bmatrix} i_a \\ i_\beta \end{bmatrix} = \begin{bmatrix} \cos\theta & -\sin\theta \\ \sin\theta & \cos\theta \end{bmatrix} \begin{bmatrix} i_d \\ i_q \end{bmatrix}$$

The model of a three-phase induction model under d-q two-axis coordinate can be expressed as follows. Votlage equations:

$$\begin{bmatrix} u_{ds} \\ u_{qs} \\ u_{dr} \\ u_{qr} \end{bmatrix} = \begin{bmatrix} r_s + L_s p & -\varpi_l L_s & L_m p & -\varpi_l L_m \\ 0 & r_s + L_s p & \varpi_l L_s & L_m p \\ L_m p & -\varpi_r L_m & r_r + L_r p & \varpi_s L_r \\ \varpi_s L_m & L_m p & \varpi_s L_r & r_r + L_r p \end{bmatrix} \cdot \begin{bmatrix} i_{ds} \\ i_{qs} \\ i_{dr} \\ i_{qr} \end{bmatrix}$$

Where, $L_s = L_{ls} + \frac{3}{2} L_{ms}, L_r = L_{lr} + \frac{3}{2} L_{ms}, L_m = \frac{3}{2} L_{ms}, p = \frac{d}{dt}; \varpi_s = \varpi_i - \varpi_r, \varpi_l$: synchronous rotating speed. Flux linkage equations:

$$\begin{bmatrix} \psi_{ds} \\ \psi_{qs} \\ \psi_{dr} \\ \psi_{qr} \end{bmatrix} = \begin{bmatrix} r_s & 0 & L_m & 0 \\ 0 & r_s & 0 & L_m \\ L_m & 0 & L_r & 0 \\ 0 & L_m & 0 & L_r \end{bmatrix} \cdot \begin{bmatrix} i_{ds} \\ i_{qs} \\ i_{dr} \\ i_{qr} \end{bmatrix}$$

Torque equation:

$$T_e = PL_m\left(i_{qs}i_{dr} - i_{ds}i_{qr}\right)$$

With these transformations, the stator current is decomposed into two DC components aligned with rotor magnetic field, id and i_q, with id corresponding to excitation current and i_q corresponding to torque current. These two currents can be controlled separately. Control of id means control the flux, while control of i_q means control of torque, which is like the control of a DC motor, and then different control strategies can be easily applied on the control of EV driven by induction motor. With vector control scheme, the structure of whole system can be given as shown in figure.

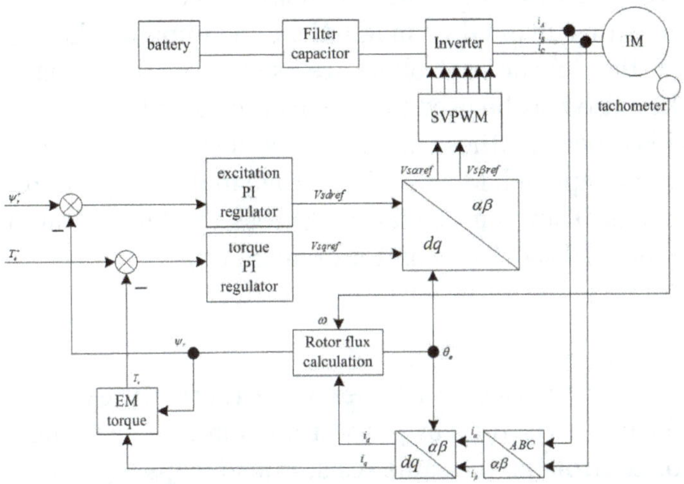

Figure: The control system of EV driven by induction motor with vector control scheme

Motor Controller

A Motor Controller is a device that acts as intermediary between your robot's microcontroller, batteries and motors. A motor controller is necessary because a microcontroller can usually only provide roughly 0.1 Amps of current whereas most actuators (DC motors, DC gear motors, servo motors etc.) require several Amps. The current a motor consumes is related to the torque it can provide (a small motor will not consume much current, but can't move a large mass whereas a large motor can move a large mass but will consume a large current).

Types of Motor Controllers

A motor controller protects a motor from damage, starts or stops it, increases or decreases its speed, or reverses its direction of rotation.

Manual

A manual (nonautomatic) controller is operated byhand directly through a mechanical system. Theoperator closes and opens the contacts that normally energize and de-energize the connected load.

Magnetic

In a magnetic controller, the contacts are closed or opened by electromechanical devices operated by local or remote master switches. Normally, all the functions of a semiautomatic magnetic controller are governed by one or more manual master switches; automatic controller functions are governed by one or more auto- matic master switches, after it has been initially energized by a manual master switch. Either controller can be operated in the semiautomatic or automatic mode, depending on the mode of operation selected.

Across-the-Line

An across-the-line controller throws the connected load directly across the main supply line. This motor controller may be either manual or magnetic, depending on the rated horsepower of the motor. Normally, across- the- line dc controllers are used for starting small (fractional horsepower) motors. However, they also may be used to start average-sized, squirrel-cage induction motors without any damage. This is because these motors can withstand the high starting currents due to starting with full-line voltage applied. Most squirrel- cage motors drive pumps, compressors, fans, lathes, and other auxiliaries. They can be started "across the line" without producing excessive line-voltage drop or mechanical shock to a motor or auxiliary.

Dc Resistor

In a dc resistor motor controller, a resistor in series with the armature circuit of the dc motor limits the amount of current during starts, thereby preventing motor damage and overloading the power system. In some resistor controllers, the same resistor also helps regulate the speed of the motor after it is started. Other dc controllers use a rheostat in the motor shunt field circuit for speed control.

AC Primary Resistor

In an ac primary resistor controller, resistors are inserted in the primary circuit of an ac motor for both starting and speed control. Some of these controllers only limit the starting currents of large motors; others control the speed of small motors, as well as limit the starting current.

AC Secondary Resistor

In an ac secondary resistor controller, resistors are inserted in the secondary circuit of a wound-rotor ac motor for starting or speed control. Although they are sometimes used to limit starting currents, secondary resistor controllers usually function to regulate the speeds of large ac motors.

Static Variable-Speed

A static variable-speed motor controller consists of solid-state and other devices that regulate motor speeds in indefinite increments through a predetermined range. Speed is controlled by either manual adjustment or actuation of a sensing device that converts a system parameter, such as temperature, into an electric signal. This signal sets the motor speed automatically.

Autotransformer

The auto transformer controller (or compensator) is an ac motor controller. It starts the motor at a reduced voltage through an autotransformer, and then it connects the motor to line voltage after the motor accelerates. There are two types of compensators: open transition and closed transition. The open-transition compensator cuts off power to the motor during the time (transition period) the motor connection is shifted from the auto- transformer to the supply line. In this short transition period, it is possible for the motor to coast and slip out of phase with the power supply. After the motor is connected directly to the supply line, the resulting transition current may be high enough to cause circuit breakers to open. The closed-transition compensator keeps the motor connected to the supply line during the entire transition period. In this method, the motor cannot slip out of phase and no high transition current can develop.

Reactor

A reactor controller inserts a reactor in the primary circuit of an ac motor during starts, and later it short.

For all DC motor brushed types here are the common reasons why,

Motor protection: Most modern controllers have the following protections; under-voltage, over-voltage, short circuit protection, current limit protection, thermal protection and voltage transients. Without these protections the motor is "exposed" to threats that will possibly result in permanent electrical or mechanical damage.

Speed control: All DC motors will lose speed as they are loaded and increase in speed when they are unloaded, in a linear fashion, according to their speed/torque gradient. For applications where a specific speed is required, with an unknown load (so a final speed cannot be calculated), or a fluctuating load (conveyor belt, pump, grinding tool, reel/converter, Cam) a controller is a must.

In the cam application in particular, where the motor is operating as a "motor", for half the cycle and a "generator" for the other half, as the load "pushes" the motor a four quadrant drive must be used. This provides dynamic breaking control, to ensure the motor remains under control and at the constant speed. It is not possible to achieve this with a simple supply or a simple single quadrant controller. The controller must also cope with the varying load, yet maintain the motor at a constant speed, by varying the voltage to the motor as the load changes to compensate.

In fan and pump applications the power curve follows the "square law" i.e. for an incremental increase of the speed the power increase is to the second power. Using a speed controller to lower the speed of the motor to meet the demands of a system, in place of constricting the flow from the pump or fan whilst running the motor at full speed will reduce the power consumption. That said there are losses across the controller, typically 1-5%, but these are minimal in comparison to the motor or the mechanics it is linked to.

Torque control: A DC motor's current is proportional to the torque delivery. Having control over the current to the motor will govern its torque delivery. Without control of the DC motor torque/current, the motor is allowed to pull large currents that can often result in torques delivered in excess of what is mechanically viable for the system, leading to imminent failure (especially when gear with a large reduction ratio is used). Also failure can be in the form of burn out from stalling the motor, for instance when the mechanics are jammed. In this condition, unless the current is limited, the motor pulls the stall current. As you can see from the diagram, the stall current is far from the operating area of the motor and will cause a thermal failure in normally few seconds.

References

- Anderson, C.D. and Anderson, J. (2005) "New Charging Systems" Electric and Hybrid Cars: a History (North Carolina: McFarland & Co., Inc.) ISBN 0-7864-1872-9, p. 121
- Eight-tips-extend-battery-life-your-electric-car-107938: plugincars.com, Retrieved 22 May 2018
- Uehara, I.; Sakai, T.; Ishikawa, H. (1997). "The state of research and development for applications of metal hydrides in Japan". Journal of Alloys and Compounds. 253: 635–641. doi:10.1016/s0925-8388(96)03090-3
- What-is-a-traction-motor: wisegeek.com, Retrieved 12 June 2018
- Russell Hensley, John Newman, and Matt Rogers (July 2012). "Battery technology charges ahead". McKinsey & Company. Archived from the original on 2017-01-09. Retrieved 2017-01-12
- Electric-traction-system: electronicshub.org, Retrieved 18 April 2018
- Taniguchi, Akihiro; Fujioka, Noriyuki; Ikoma, Munehisa; Ohta, Akira (2001). "Development of nickel/ metal-hydride batteries for EVs and HEVs". Journal of Power Sources. 100: 117–124. doi:10.1016/s0378-7753(01)00889-8
- What-is-a-motor-controller-6905: robotshop.com, Retrieved 28 June 2018
- "Electric Car Battery Prices on Track to Drop 70% by 2015, Says Energy Secretary : TreeHugger". treehugger.com. Retrieved 2014-02-01

5

Electric Vehicle Charging Devices and Systems

An electric vehicle charging station is an infrastructure that supplies electric power for recharging electric vehicles, such as plug-in hybrids, electric cars, etc. This chapter has been carefully written to provide an easy understanding of the varied aspects of electric vehicle charging devices and systems, with a compilation of significant topics such as charging station, connectors, rapid chargers, fast chargers, slow chargers, wireless charging, EnergyBus connectors, etc.

Charging Station

An electric vehicle charging station is equipment that connects an electric vehicle (EV) to a source of electricity to recharge electric cars, neighborhood electric vehicles and plug-in hybrids. Some charging stations have advanced features such as smart metering, cellular capability and network connectivity, while others are more basic.

Charging stations are also called electric vehicle supply equipment (EVSE) and are provided in municipal parking locations by electric utility companies or at retail shopping centers by private companies. These stations provide special connectors that conform to the variety of electric charging connector standards.

Fees for using EVSE vary from monthly or yearly flat rates to per-kWh to hourly rates. Charging stations can be free and are usually subsidized by the local government.

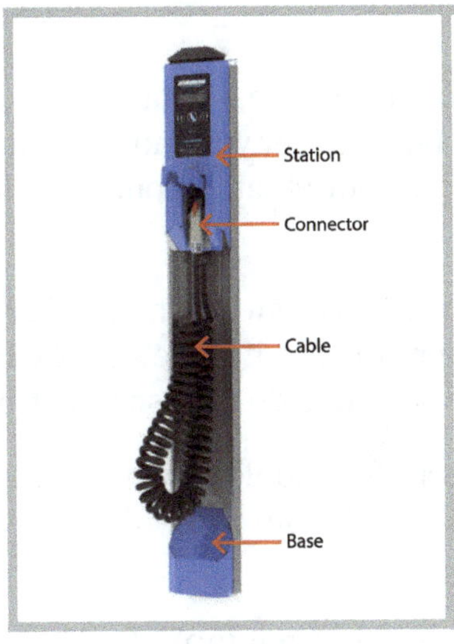

Charging station

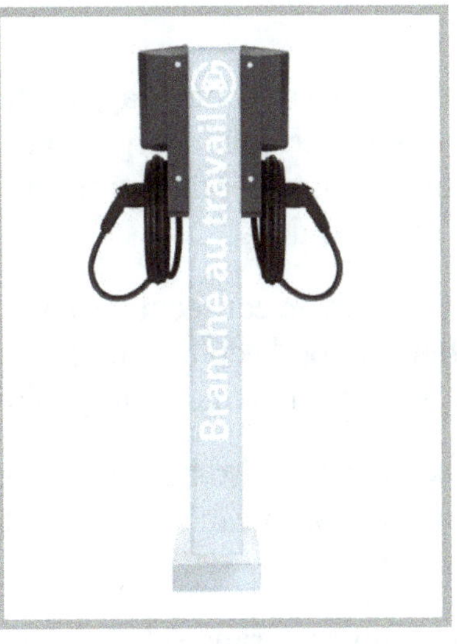

Doul charging station

A charging station is usually in the form of a fixture connected directly to an electrical distribution panel, or sometimes to an electrical outlet. It has one or more charging cables equipped with a connector that is similar to a gas-pump nozzle and is used in the same way: it simply connects to the EV's charging socket to charge the battery. The station has lights that indicate that the EV is connected and charging. It can also have a button for starting or stopping the charging operation. Some have additional features: energy meter, electronic payment system, card-controlled access system, Internet access, etc.

Different types of EVSE Provide different Speeds of Charging. It is described in 2 Levels.

Level 1 Charging

Level 1 Charging – Standard House Outlet

Level 1 charging is the technical jargon for plugging your car into an ordinary household outlet. For a Leaf, this means about 4.5 miles of range per hour of charging, or about 22 hours for a full charge. Wow, does that sound terrible! But there's a problem with thinking this way: you'll rarely need to do a full charge from flat empty to full. If you drive 40 miles per day and charge overnight, you'll be back to full in 9 hours. When you're sleeping, it doesn't matter if it takes one hour or 9 hours to charge.

But what if you have to drive a lot one day, say 80 miles? Sure, it would take 18 hours to get a full charge, but with a 9-hour overnight charge, you'll be ready for your normal commute the next day. If you drive less than 40 miles per day or charge for more than 9 hours, you'll work back up to a full charge over the next few days.

If you need to drive 80 miles on consecutive days, you'll need an alternative. Maybe you'll drive your other car, that gas-burner you keep around for long trips, or if there's public EV charging in your area, you can charge away from home while you're parked to do your shopping or other errands.

Level 1 charging at work could also be a supplement for people driving over 40 miles per day, or even a substitute for those who can't charge at home (because they don't have a garage or fixed parking place, for example).

Since it's easy to get 40 miles of range charging overnight from 120V, Level 1 is perfectly suited for overnight charging of the Chevy Volt, a plug-in hybrid with a 40-mile all-electric range.

Although Level 1 charging is generally too slow for a road trip, it can be helpful as destination charging. Cathy and I drove 90 miles to San Juan Island, charged for a few days in a friend's garage when not cruising around the island, and left with a full charge. That was great, but I wouldn't want to have to wait for Level 1 charging in the middle of a travel segment.

Beyond range issues, Level 1 may not be suitable for primary charging in all cases. In extreme climates, more power may be required to maintain proper battery temperatures. In these cases, Level 2 charging may be more appropriate.

DC Fast Charging

Blink DC Fast Charge Station

At the other end of the spectrum is DC Fast Charging, the fastest type of charging currently available. It provides up to 40 miles of range for every 10 minutes of charging. These stations are expensive (up to $100,000) and require more power than your house, so you'll never have one of these in your garage.

They are going to start appearing as public charging stations in the next year, beginning in the Leaf target areas. If there's one conveniently located near where you drive, you can get back up to 80% of a full charge while getting lunch or drinking a latte. Charging this fast makes it far more practical to drive beyond an EV's single-charge range in one day. It's still not going to make a one-day 800-mile drive practical, but a 200-mile drive with a couple of charging breaks can be quite doable.

Level 2 Charging

Between the cheap Level 1 and expensive DC Fast Charging stations sits Level 2 charging. Level 2 supplies 240V, like what an electric dryer or oven uses. It goes through a box and a cord that improves safety by waiting to send power to the plug until it's plugged into an EV. Level 2 allows for a wide range of charging speeds, all the way up to 19.2 kilowatts (kW), or about 70 miles of range per hour of charging.

ChargePoint/Coulomb Level 2 Charging Station

However, the charging stations being put in with federal grant money don't support the full range of Level 2 charging and max out at 6.6 kW or around 26 miles of range per hour of charging.

Both Level 1 and Level 2 charging stations simply deliver household electricity to the car. Electronics on board the car transform the wall power into the proper form to charge the battery. This bit of electronics built into the car also has a maximum power rating. The first model-year Leafs can only use 3.3 kW, about 12 miles of range per hour, or about 8 hours for a full charge from empty. The Chevy Volt's on-board charger is also limited to 3.3 kW, although its smaller battery pack gets full sooner.

Nissan recommends that you install a Level 2 charging station at home. That's a reasonable thing to do if you don't mind spending about $2,000, just consider it part of the cost of the car. Early buyers in the Leaf target markets may be able to get into The EV Project and get a free Level 2 charging station plus an allowance toward the install cost. Failing that, there's a 30% federal tax credit (up to $1,000) for installing EV charging, which can make it less expensive. Still, if you are planning to use your EV for a daily commute of 40 miles or less per day, you should at least consider using Level 1 charging at home. You can always add a Level 2 charging station later if you decide you need it.

There will soon be 20,000 public Level 2 charging stations (limited to 6.6 kW) installed mainly in the Leaf target areas. Even if you only have Level 1 charging in your garage, if you're in the early rollout areas, you should have access to convenient Level 2 charging available while your car is parked and you're doing something else. These charging stations will make it possible to drive 60 miles to a baseball game and pick up about 50 miles of range in 4 hours while you're having fun, thus easily driving over the single-charge range while always keeping a healthy reserve.

Charge Time and Battery Capacity

It's misleading that charging times are generally quoted as time for a full charge. While it does take about 22 hours (Level 1) or 8 hours (Level 2) to charge a Leaf from empty to full, you're not likely to do that often because you will rarely arrive home with a fully depleted battery. It doesn't matter if you're driving a 40-mile Volt, a 100-mile Leaf or a 240-mile Tesla Roadster, if your commute is 40 miles, you'll only need about 9 hours (Level 1) or 3 hours (3.3 kW Level 2) to charge.

Cost of Charging an Electric Car at a Public Station

When you drive your EV into the wild and look for a charge, the situation gets a bit more complicated. Electric car owners often hold multiple charging station provider accounts to cover various needs when traveling.

If you only drive your EV in and around the city, a ChargePoint account is likely to cover your daily needs with Level 2 plugs. While membership is free, ChargePoint has users pay in $25 increments. The stations have different price levels set by the actual property owner or lessee.

In the course of your travels, you may see pricing set at $1 per hour up to $5 per hour for Level 2 stations. Some stations are also free. These prices do not include the cost of access to the parking facility, which depends on the location.

Electric vehicle drivers who hope to save time with DC fast-charging will usually pay more for the convenience. A popular provider in this department is NRG EVgo, which has plans available in Los Angeles for $15 per month plus $0.10 per minute for fast-charging sessions ($1 per hour for Level 2 charging). Other plans come without a monthly fee and cost $11 for 30 minutes of fast-charging.

Magna-Charge System

The Magna-Charge system consists of two parts:

- A charging station mounted to the wall of the house.
- A charging system in the trunk of the car.

The charging station is hard-wired to a 240-volt 40-amp circuit through the house's circuit panel.

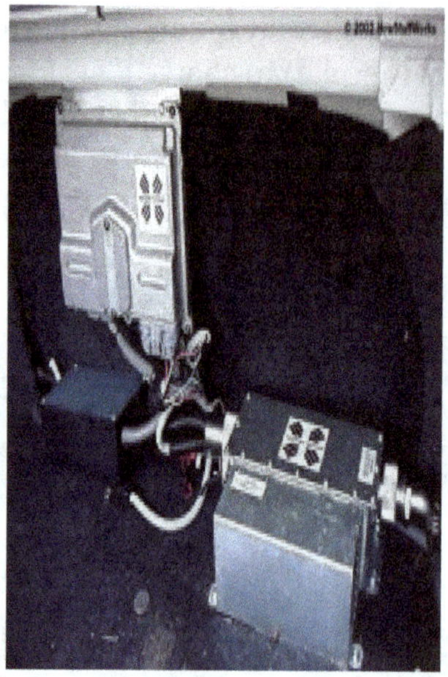

The charging system sends electricity to the car using this inductive paddle:

The paddle fits into a slot hidden behind the license plate of the car.

The paddle acts as one half of a transformer. The other half is inside the car, positioned around the slot behind the license plate. When you insert the paddle, it forms a complete transformer with the slot, and power transfers to the car.

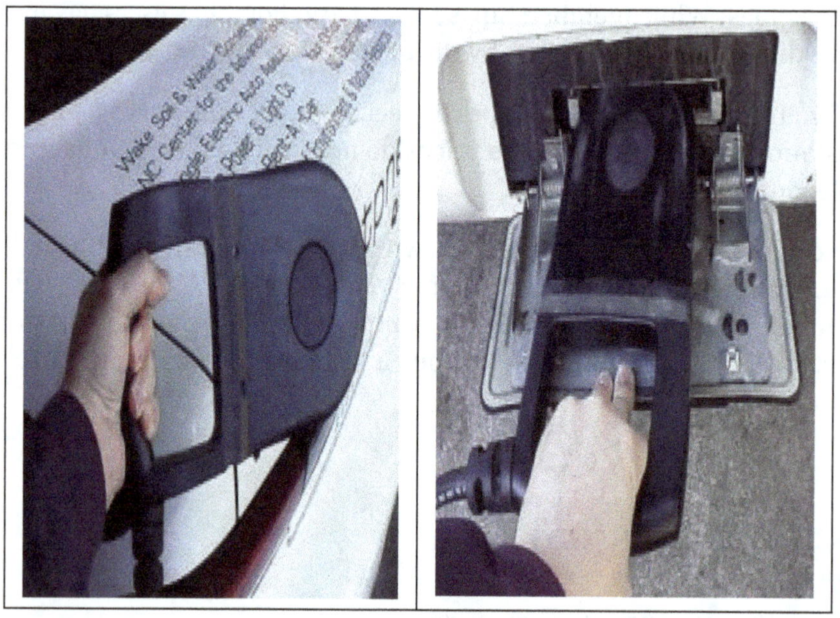

One advantage of the inductive system is that there are no exposed electrical contacts. You can touch the paddle or drop the paddle into a puddle of water and there is no hazard. The other advantage is the ability to pump a significant amount of current into the car very quickly because the charging station is hard-wired to a dedicated 240-volt circuit.

The inductive paddle are divided into two Parts: Large Paddle Inductive and Small Paddle Inductive.

- Large Paddle Inductive (LPI) or first-gen Magne Charge is an inductive charging system used in conjunction with battery powered electric vehicles (BEV) formerly made by General Motors. LPI is the precursor to SPI or Small Paddle Inductive and is not compatible with Small Paddle Inductive vehicles.

  Magne Charge support was withdrawn by General Motors in 2002, after the California Air Resources Board settled on a conductive charging interface for electric vehicles in California in June, 2001. It has also been referred to as Magne-Charge and MagneCharge.

- Small Paddle Inductive (SPI) or second-gen Magne Charge is an inductive charging system used in conjunction with battery powered electric vehicles (BEV) formerly made by General Motors. SPI it is the follow up to LPI or Large Paddle Inductive and is not compatible with Large Paddle Inductive vehicles.

  Magne Charge support was withdrawn by General Motors in 2002, after the California Air Resources Board settled on a conductive charging interface for electric vehicles in California in June, 2001. It has also been referred to as Magne-Charge and MagneCharge.

Connectors

The choice of connectors depends on the charger type (socket) and the vehicle's inlet port and charging standard. On the charger-side, rapid chargers use CHAdeMO, CCS (Combined Charging Standard) or Type 2 connectors, Fast and Slow units.

In the electric vehicle market there is no standardization in the connectors, users find different models and brands, with different configurations and technical characteristics. For this reason, we should keep informed about the compatibilities that exist and the different types of connectors that we will find on public roads or to install in our homes. The different types of connectors are:

- Connector for electric vehicle: Schuko plug (conventional plug). It is compatible with European outlets and meets the CEE 7/4 Type F standard. It has a ground connection, two terminals and supports current of up to 16 amps, so it is only compatible with slow recharges. It is common in some motorcycles and electric bicycles, even in some electric car like the Twizy.
- Connector for electric vehicle: Connector SAE J1772 (Type 1). It is the standard connector in Japan and has been accepted in the European Union and in the American market. It has 5 terminals, two of them current, another two complementary and the last one is the ground. This type of connector has two levels, one of them up to 16 Amps, which would be for slow recharging. The other level is up to 80 Amperes, which corresponds to fast recharge. This connector is for recharging electric vehicles such as the Opel Ampera, Nissan Leaf, Nissan ENV200, Mitsubishi Outlander, Mitsubishi iMiev, Peugeot iON, Citroen C-Zero, Renault Kangoo ZE (type 1), Ford Focuselectric, Toyota Prius Plug in, or the KIA SOUL EV.

- Connector for electric vehicle: Connector MENNEKES (Type 2). It is not specific for electric vehicles but its use is very common. It is of German origin, has 7 terminals, of which 4 are for current (three phase), another ground and two for communications. In this type we have the option of two currents, on the one hand monophasic (up to 16 A- slow recharge), and on the other three-phase (up to 63 A / 43,8kW-fast recharge). Suitable for models such as the BMW i3, i8, BYD E6, Renault Zoe, Tesla Model S, Volvo V60 plug-in hybrid, VW Golf plug-in hybrid, VW E-up, Audi A3 E-tron, Mercedes S500 plug-in, Porsche Panamera, or the Renault Kangoo ZE.
- Connector for electric vehicle: Combined single connector or CCS. Created by Germans and North Americans as a standard solution. It has 5 terminals distributed for current, grounding and communication with the network. This type of connector supports both reloads (slow and fast). Manufacturers such as Audi, BMW, Daimler, Porsche and Volkswagen already incorporate this type of connector.
- Conector para vehículo eléctrico: Conector Scame (Tipo 3): Tiene 5 o 7 bornes, dependiendo si la corriente monofásica o trifásica, incluyendo en ambas tierra y comunicación con la red. Admite hasta 32 A, y es para recarga semi-rápida.
- Connector for electric vehicle: Scame connector (Type 3): It has 5 or 7 terminals, depending on the single-phase or three-phase current, including on both earth and communication with the network. It admits up to 32 A, and is for semi-fast recharge.
- Connector for electric vehicle: Connector CHAdeMO Connector: Developed by the Japanese association formed by the Tokyo Electric Power Company (TEPCO), Nissan, Mitsubishi, Fuji Heavy Industries (Subaru) and Toyota. Designed for fast recharging in direct current, therefore it has 10 terminals, grounding and communication. This connector is for ultra-fast recharges, as it supports up to 200 A. It is equipped for cars such as the Nissan Leaf, Nissan ENV200, Mitsubishi Outlander, Mitsubishi iMiev, Peugeot iON, Citroen CZero, or the KIA SOUL EV.

Rapid Chargers

- 50kW DC charging on one of two connector types;
- 43kW DC charging on one connector type;

- 120kW DC charging on Tesla Supercharger network;
- All rapid units have tethered cables.

Rapid chargers are the fastest way to charge an EV, often found in motorway services or in locations close to main roads. Units supply high power direct or alternating current – AC or DC – to recharge a car to 80% in 20-40 minutes. All come with the charging cable tethered to the unit, with one of three connectors attached.

Rapid charging can only be used on vehicles with rapid-charging capability. With two of the three different connector types available being rapid-charging specific.Rapid DC chargers provide power at up to 50kW (125A), use either the CHAdeMO or CCS charging standards, and are indicated by purple icons on Zap-Map.

Both connectors are typically able to charge an EV to 80% in 20-40 minutes depending on battery capacity and starting state of charge. Once charging reaches 80%, the unit's power output will drop to a slower rate to preserve battery life and maximise charging levels.

Users of rapid DC units select the appropriate connector for their vehicle and use the tethered cable to plug the car in, rather than their own cable.

Tesla's Supercharger network also provides Rapid DC charging to drivers of its cars, but at a much higher rate of up to 120kW. Like other Rapid DC units, the cable is tethered to the unit, but the connector at the end is Tesla's version of Type 2.

While all Tesla models are designed for use with Supercharger units, many Tesla owners use adaptors which enable them to use a 50kW rapid units fitted with a CHAdeMO connector. While these provide less power than a Supercharger, they are far more common in the UK and elsewhere.

Rapid AC chargers provide power at up to 43kW (three-phase, 63A) and use the Type 2 charging standard. They are indicated by green icons on Zap-Map.

Rapid AC units are typically able to charge an EV to 80% in 20-40 minutes depending the model's battery capacity and starting state of charge. Once charging reaches 80%, the unit's power output will drop to a slower rate to preserve battery life and maximise charging levels.

Users of rapid AC units select the Type 2 connector for their vehicle and use the tethered cable to plug the car in, rather than their own cable.

Rapid Connectors

CHAdeMO – 50kW DC	CCS – 50kW DC

Type 2 – 43kW AC	**Tesla Type 2- 120kW DC**
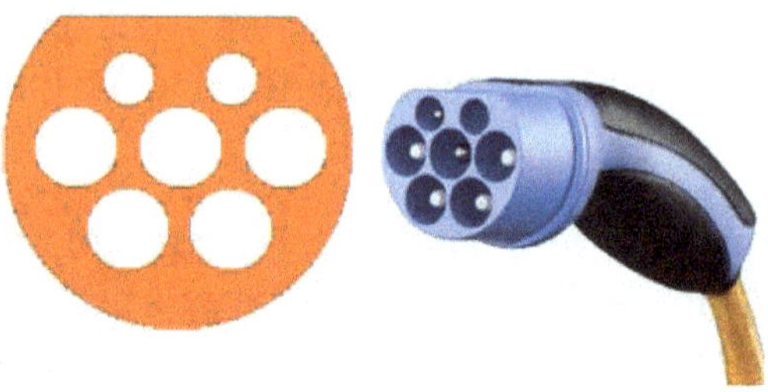	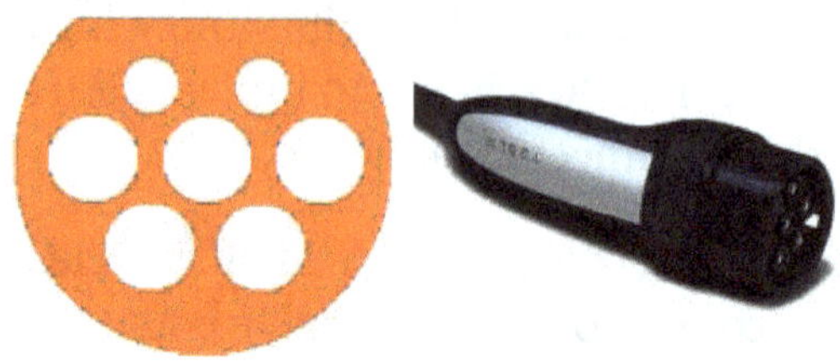

EV models that use CHAdeMO rapid charging include the Nissan Leaf, Mitsubishi Outlander PHEV, and Kia Soul EV. CCS compatible models include the BMW i3, VW e-Golf, and Hyundai Ioniq Electric.

Fast Chargers

- 7kW fast charging on one of three connector types.
- 22kW fast charging on one of three connector types.
- 11kW fast charging on Tesla Destination network.

- Units are either untethered or have tethered cables.

Fast chargers, all of which are AC, are typically rated at either 7kW or 22 kW (single- or three-phase 32A). Charging times vary on unit speed and the vehicle, but a 7kW charger will recharge a compatible EV in 3-5 hours, and a 22kW charger in 1-2 hours.

These type of chargers tend to be found at destinations, such as car parks, supermarkets, or shopping centres – somewhere that an EV will potentially be parked at for an hour or more.

The majority of fast chargers are untethered, though some home and workplace based units have cables attached. The latter units mean only those vehicles that can use that connector type will be able to charge on them; in contrast to the more common use of a driver's own connector cable. Untethered units are therefore more flexible and can be used by any driver that has the correct cable.

The most common type of fast charger is an untethered 7kW Type 2, though fast chargers can also be found with Type 1 or Commando connectors. Fast charge units commonly have two sockets to charge two cars at the same time, though one is not unusual.

Charging speeds from fast chargers will depend on the car's on-board charger, with not all models able to accept 7kW or more. These models can still be plugged in to the charge point, but will only draw the maximum power accepted by the on-board charger. For example, a Nissan Leaf with standard 3.3kW on-board charger will only draw a maximum of 3.3kW, even if the fast charger is 7kW or 22kW.

Type 1 – 7kW AC	Commando – 7-22kW AC
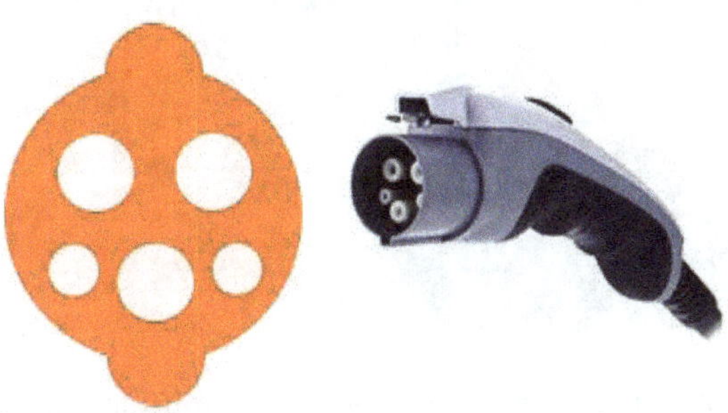	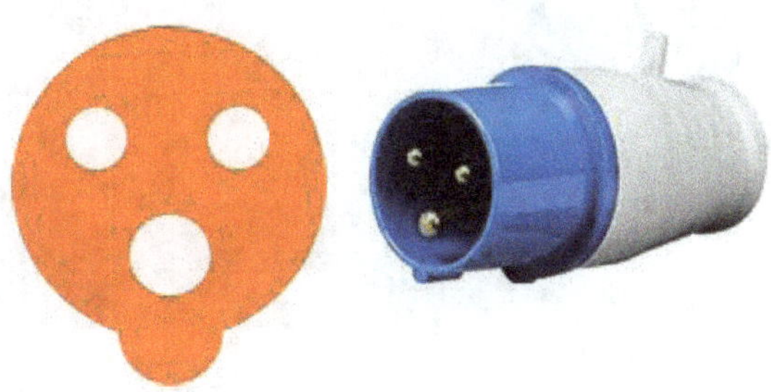

Almost all EVs and PHEVs are able to charge at Type 2 units, with the correct cable at least. It is by far and away the most common public charge point standard around, and most plug-in car owners will have a cable with a Type 2 connector charger-side.

There are some charge points around that have use a Commando or Type 1 connector, but these are relatively few and far between, with the dominant standard being Type 2.

Example

Tesla's Destination Chargers provide 7, 11, or 22kW of power but, like the Supercharger network, can only be used by Tesla models. Tesla does provide some standard Type 2 chargers at many of its destination locations, and these are compatible with any plug-in model using the correct cable.

Slow Chargers

- 3kW slow charging on one of four connector types.
- Charging units are either untethered or have tethered cables.
- Includes mains charging and from specialist chargers.
- Often covers home charging.

Slow charging units are rated at 3kW. Charging times vary on unit speed and vehicle, but a full charge for an EV will typically take 6-12 hours.

Slow charging is the most common method of charging electric vehicles, used by many owners to charge at home overnight. Slow units aren't necessarily restricted to home use, with workplace and public points also able to be found. Because of the longer charging times over fast units, slow public charge points are less common.

Home charge points are commonplace though, largely because those who buy an electric car often find themselves qualified to apply for the Electric Vehicle Homecharge Scheme. This gives buyers money off a fully installed EV home charging unit.

Slow charging units can be either tethered or untethered, with untethered charge points often using a Type 2 inlet to connect an EV with. Tethered points typically have a Type 1 connector where this type is required by an owners' EV model.

Although termed 3kW units, slow home charge points can actually potentially charge at up to 3.6kW, because they can be rated for 16A, rather than the 13A or less available from the mains.

While slow charging can be carried out via a three-pin socket too using a standard 3-pin socket, because of the higher current demands of EVs and the longer amount of time spent charging, it is strongly recommended that those who need to charge regularly at home or the workplace get a dedicated EV charging unit installed by an accredited installer.

Slow Connectors

3-Pin – 3kW AC

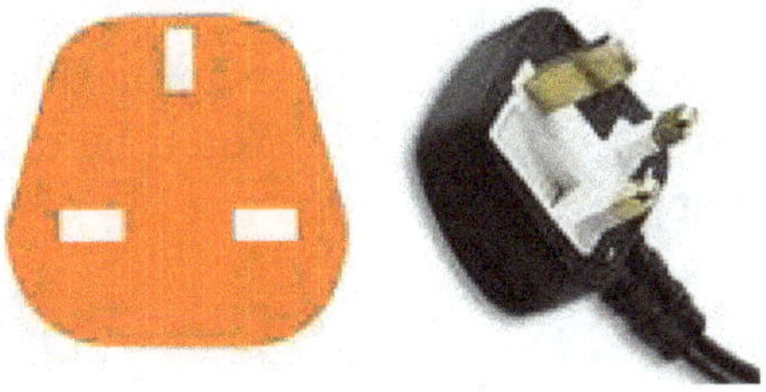

All plug-in electric vehicles can charge using at least one of the above slow connectors. Often home charge points will be compatible with the same Type 2 cable used for public charging, or be tethered with the a Type 1 connector where this is suitable for a particular EV.

Combined Charging System

The Combined Charging System (CCS) is a combined charging system for charging electric vehicles with alternating current (AC) and direct current (DC). Both AC and DC Vehicle Connectors can be charged via the Vehicle Inlet, which means that only one interface is required in the vehicle.

- Conventional AC charging overnight in the garage or in parking lots when parked for longer periods of time
- Fast DC charging in a few minutes at rest stops, restaurants, supermarkets, banks, etc.

The Combined Charging System (CCS) is standardized in IEC 62196-3 and SAE J1772. Connector geometries are available for Type 1 (North America) and Type 2 (Europe). The system was developed in collaboration with leading automobile manufacturers and is already used in a large number of electric vehicles.

Starting a Charge Session

- Firstly, make sure your vehicle is fully switched off.
- The electricity pump screen will show "Welcome! To charge your car use the Electric Highway smartphone app".
- Then simply use the EV charging app to begin charging. You'll need to follow the instructions until you are prompted to scan the QR codes (or enter the location ID manually) found on the front of the charging pump.
- Using the app, select which connector to use to charge your vehicle.
- You'll then be prompted to confirm the stored card to use to pay for the charging session.
- Follow the instructions on the charging pump screen, to see how to connect your vehicle.
- There's no mechanism to lock the cable into place, so just ensure that the connector is pushed in firmly.
- The charge point will start the charge session once it has confirmed that the vehicle is properly connected.
- After checking your vehicle's dashboard (where applicable) to confirm charging is in progress, you're then able to leave your vehicle. You'll be able to view the progress of your charging session on your app.

Ending a Charge Session

- Once the charge has ended, "Your charge session is complete" will be shown on your app screen. At this point you can return to your vehicle to disconnect the charging cable.
- Follow the instructions on the charging pump screen to disconnect your vehicle. Simply hang the charging cable back up and press the button below and to the right of the green tick.
- You can then use the app to indicate that you've disconnected from the pump. You'll see a screen summarising your payment.
- You'll then be emailed a receipt.

If you want to Finish your Charging Session Early

- On your app, select 'Stop session early'.
- You'll be prompted to check that you're sure you want to finish early, as you still need to pay for the session you have used so far.
- Once you've confirmed finishing your session early, you can then follow the instructions on the charging pump screen to disconnect your vehicle.

Advantages

- Maximum possible flexibility when charging the electric vehicle
- Fast interim charging as well as charging overnight in the garage
- Safe charging with the help of actuator locking and temperature monitoring of the power contacts
- One charging system for the whole of Europe and the whole of North America

Actuator Locking

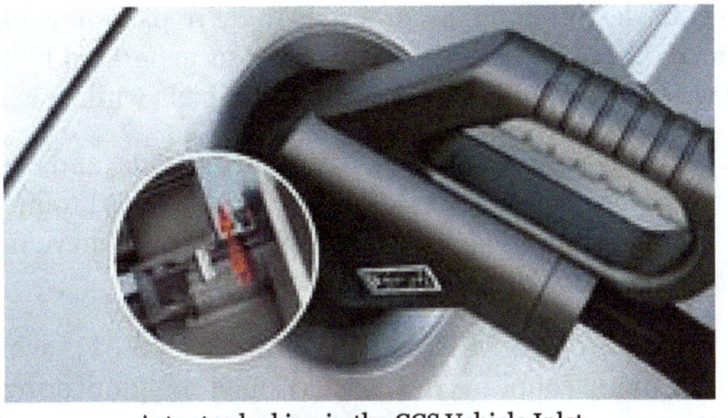

Actuator locking in the CCS Vehicle Inlet

As specified in the standard, the CCS Vehicle Inlets are equipped with an electromechanical locking actuator. It locks the Vehicle Connector to the side or directly to the locking clip in the pin connector pattern during the charging process. The actuator bolt is designed to withstand high pull-out forces.

Temperature Monitoring

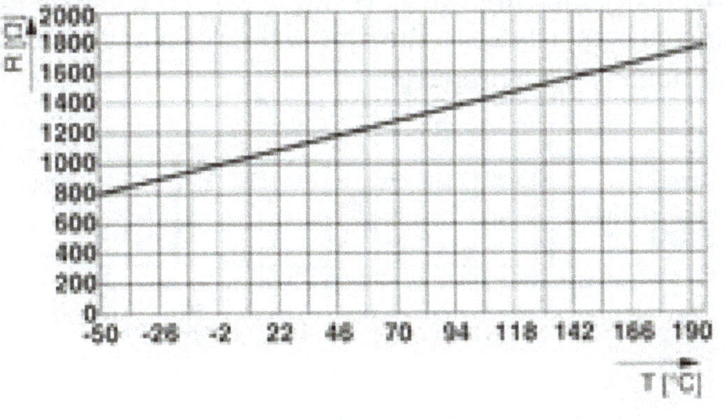

Characteristic curve of the PT1000 temperature sensor

Monitoring the charging current is also necessary for a safe charging process. If the system should overheat, e.g., in the event of high outside temperatures or an overload, this is detected by the temperature-dependent PT1000 resistance temperature detectors. The temperature of the power contacts is communicated to the charging station via the corresponding signal outputs.

In the event of overheating, the charging station is then able to switch off the charging process or reduce the charging power. As per IEC 62196, the temperature increase may not exceed 105°C.

Wireless Charging

The successful convergence of new technologies will require electric vehicles (EVs) that are low cost and fully autonomous. These attributes can be realized through wireless charging.

Consider a future in which a driverless ridesharing electric vehicle (EV) pulls over as you exit a building, takes you to your destination, and proceeds to drive passenger after passenger without ever needing to stop to recharge its battery. Instead, power generated by nearby wind and solar resources is delivered wirelessly from the roadway to the vehicle while it is in motion.

Not having to stop for recharging will make EVs truly autonomous, and, because the vehicles can thus remain in service for more hours, fewer vehicles will be needed to meet passenger demand. Furthermore, EVs with in-motion (dynamic) wireless charging can have much smaller batteries, an option that can reduce their cost and accelerate adoption.

While the concept of medium-range wireless power transfer (WPT), achieved using near-field (non-radiative) electromagnetic coupling, has existed since the pioneering work of Nikola Tesla (1891) more than a century ago, the technology to enable effective dynamic WPT for EVs is still in its nascent stage. Numerous challenges related to performance, cost, and safety need to be overcome before the vision of wirelessly powered EVs can be realized.

Near-field Wireless Power Transfer

Near-field WPT systems are of two types: inductive, which use magnetic field coupling between conducting coils, and capacitive, which use electric field coupling between conducting plates to transfer energy. For medium-range applications (in which the distance between the transmitter and the receiver couplers is comparable to the size of the couplers, as in EV charging), inductive WPT systems have traditionally been preferred.

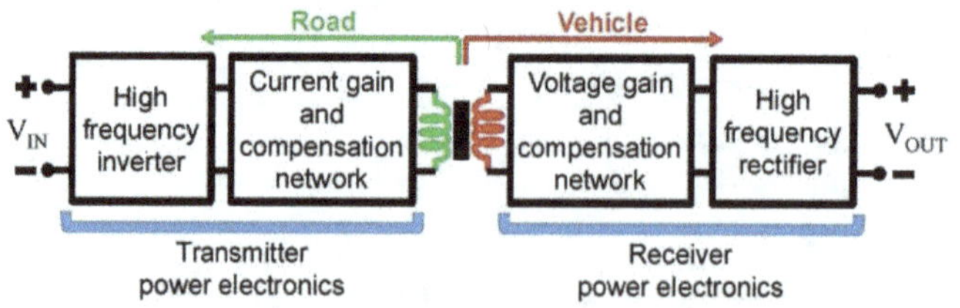

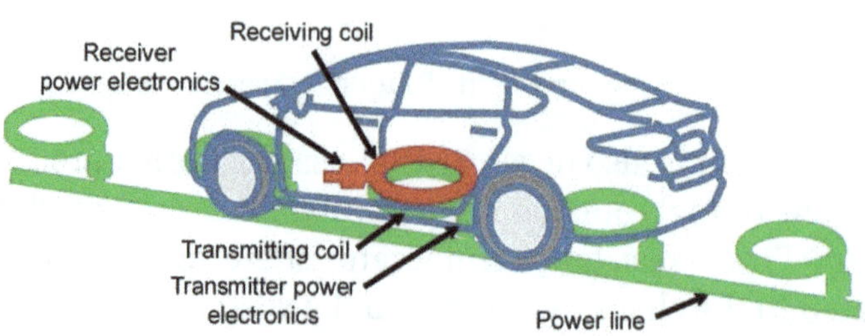

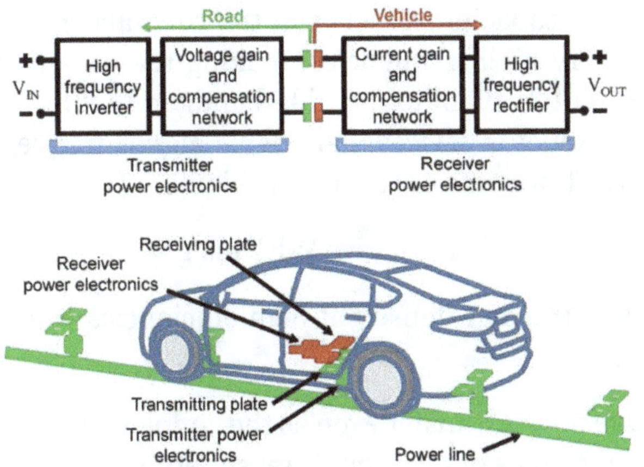

Physical implementation (left) and block diagram representation (right) of two approaches to deliver energy wirelessly to electric vehicles from an electrified roadway:

(a) Inductive wireless power transfer (WPT) using coils (embedded in the roadway and in the vehicle) that are coupled through magnetic fields,

(b) Capacitive WPT using plates coupled through electric fields. In both cases, power electronics (comprising a high-frequency inverter and rectifier with semiconductor devices, and gain and compensation networks with inductors, capacitors, and transformers) is the enabling technology.

Inductive WPT Systems

Building on work done for material handling applications during the 1990s, the past decade has seen tremendous progress in inductive WPT technology for stationary charging of EVs. Aftermarket stationary chargers are already available, and some EV manufacturers have announced plans to introduce built-in stationary inductive WPT systems as early as 2018.

However, for magnetic flux guidance and shielding, inductive WPT systems require ferrite cores, making them expensive and bulky. Also, to limit losses in the ferrites, the operating frequencies of these systems are kept under 100 kHz, resulting in large coils and low power transfer densities. The high cost and low power transfer density are particularly problematic for dynamic WPT, as these systems need to have very high power capability to deliver sufficient energy to the vehicle during its very brief time passing over a charging coil.

For these reasons dynamic inductive WPT is yet to become commercially viable, although a few experimental systems have been demonstrated.

Capacitive WPT Systems

Capacitive WPT systems have potential advantages over the inductive systems because of the relatively directed nature of electric fields, which reduces the need for electromagnetic field shielding. Also, because capacitive WPT systems do not use ferrites, they can be operated at higher frequencies, allowing them to be smaller and less expensive. Capacitive WPT could thus make dynamic EV charging a reality.

But because of the very small capacitance between the road and vehicle plates, effective power transfer can occur only at very high frequencies, making the design of these systems extremely challenging. With the recent availability of wide-bandgap (gallium nitride [GaN] and silicon carbide [SiC]) power semiconductor devices that enable higher-frequency operation, high-power medium-range capacitive WPT systems are becoming viable.

Two major challenges associated with capacitive WPT for EV charging are:

(1) Achieving high-power transfer density at high efficiencies while meeting electromagnetic safety requirements,

(2) Maintaining effective power transfer even as the couplers' relative position changes. These challenges have been a focus of my group's recent efforts.

Achieving Safe and Efficient High-power Transfer

The size of the couplers in WPT systems can be reduced and the power transfer density increased by designing the systems to operate at higher frequencies. In inductive systems the increase in induced voltage with higher frequency compensates for the reduced mutual inductance of the smaller coils, and in capacitive systems the increase in displacement current with higher frequency compensates for the smaller plates' lower capacitance. Higher operating frequencies also enable smaller power electronics associated with WPT systems thanks to a decrease in energy storage requirements.

But achieving high efficiencies at high switching frequencies is very challenging. And the fringing fields of WPT systems must be within safe levels in areas occupied by people and animals (e.g., the vehicle cabin and outside the perimeter of the chassis). These requirements for capacitive WPT systems can be met through circuit stages that provide appropriate voltage and current gain (to reduce displacement currents) as well as reactive compensation. An active area of research is the design of these circuit stages.

Phased-array Field Focusing

Phased-array field focusing provides opportunities for innovation, for example in the exploration of methods that incorporate parasitic interactions between multiple coupling plates in the design of the matching networks. Such phased-array approaches could also be adapted for inductive WPT to help eliminate ferrites.

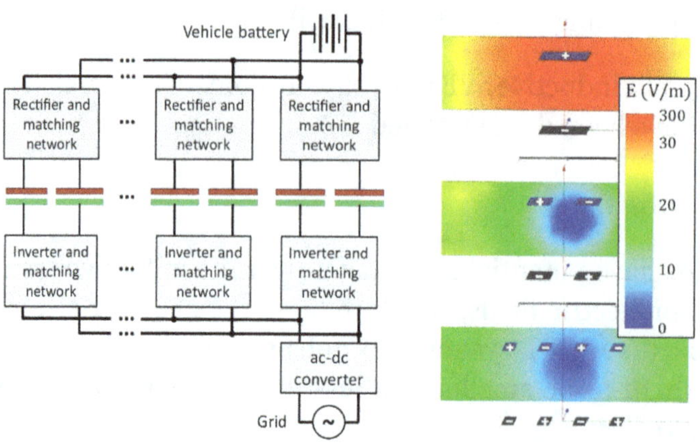

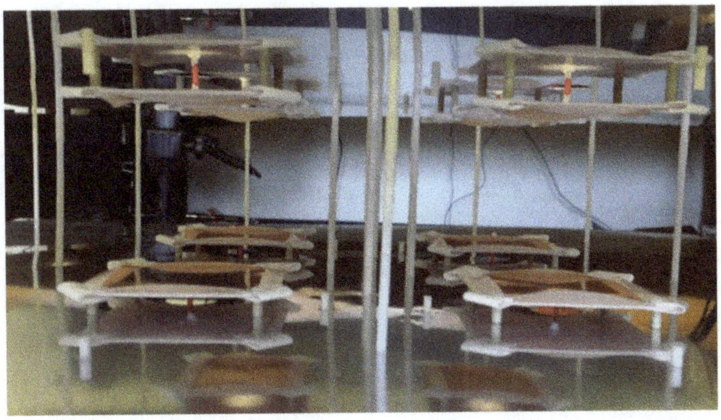

Multi-modular near-field phased-array capacitive wireless power transfer (WPT) system:

(a) Block diagram representation,

(b) Simulated performance showing fringing field reduction with progressive increase in the number of modules,

(c) Photograph of a prototype system. E = electric field strength; V/m = volt per meter.

Achieving Variable Compensation

To achieve effective power transfer, WPT systems need to operate close to the resonant frequency of the resonant tank formed by the reactance (capacitive and inductive) of the coupler and compensating network. However, the coupler reactance depends on the vehicle's road clearance, and varies as the vehicle moves across the charger. The drift between resonant and operating frequency causes a reduction in power transfer and WPT system efficiency.

In WPT systems that operate at frequencies below 100 kHz, where bandwidths are not restrictive, the traditional way to deal with variations in coupling is to change the operating frequency to track the resonant frequency. But in high-frequency WPT systems the operating frequency must stay within one of the designated, very restrictive industrial, scientific, and medical (ISM) bands.

One solution, employed in low-power inductive WPT systems, is to use a bank of capacitors that can be switched in and out of the compensating network, to keep the resonant frequency roughly unchanged as the transmitter and receiver move relative to each other. But this is not an effective approach for higher-power WPT systems as the switches have to be much bigger and more expensive to keep the system efficient. This approach is also less suited to capacitive WPT because it requires multiple switchable compensating inductors, which are bigger than capacitors.

Other adaptive impedance matching techniques include the use of saturable and variable inductors, but these techniques reduce system efficiency and do not scale well with power. We have developed new high-frequency rectifier and inverter architectures that compensate for coupling variations while operating at fixed frequency and maintaining high efficiency. An example is the active variable reactance (AVR) rectifier. By appropriately controlling the output voltages of its two coupled rectifiers, the AVR can provide continuously variable compensation while maintaining optimum soft switching to ensure high efficiency. This compensation architecture ensures that the output power of the WPT system is maintained at a fixed level across wide variations in coupling and is applicable to both capacitive and inductive WPT systems.

(a)

(b)

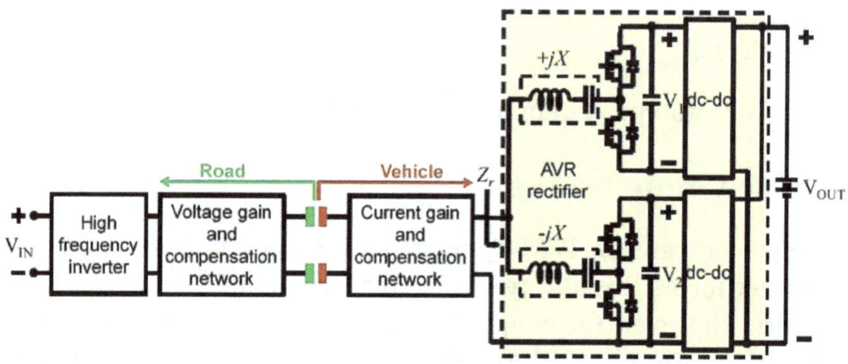

Coupling variations and an approach to compensate for these variations: (a) variation in coupling due to different vehicle road clearances, (b) variation in coupling due to change in vehicle position as it drives over the charger, and (c) a capacitive wireless power transfer (WPT) system with an active variable reactance (AVR) rectifier that can provide continuously variable compensation by controlling the voltages V1 and V2. jX = tank reactance; Zr = rectifier input impedance.

EnergyBus Connectors

EnergyBus Connector is an open standard for all electrical components of Light-Electric-Vehicles (LEVs).

It aims to improve compatibility and safety of LEVs and consists of a cutting-edge conector family and the respective communication protocol.

EnergyBus consists of a standardized set of connectors, which safely connects electric components of light electric vehicles, such as batteries, chargers, motors, sensors, human interface. These components communicate with each other on a CAN-based protocol. The uniform exchange of information enables the components to provide maximum performance. What makes EnergyBus special, apart form being the only existing standard for LEVs worldwide, is that energy AND service data can safely be transmitted through the same connector at the same time.

Goals

- Facilitation: EnergyBus will solve many problems that the LEV industry is facing as it grows, i.e. the compatibility of electric components and safety.
- Innovation: Compatible components will make product development more time and cost-efficient and contribute to a larger choice of high-quaity vehicles on the market.
- Interoperation: EnergyBus will be the basis to design public LEV infrastructure such as rental stations, or swapping machines without the monopoly of a single manufacturer.

General Benefits

EnergyBus will bring many significant benefits to everyone who develops, produces, sells, or uses LEVs or their electric components.

Electric Components will be Compatible and Safer

- Standardized connectors and communication protocol will connect components of different voltages from different manufacturers without causing damage.

- New technologies will be compatible to existing systems.
- Spare parts will be available even for older products because they can easily be exchanged.
- The bicycle industry not only works with a lot of standards for construction, production, and service; it also operates with compatible components. EnergyBus will elevate electric systems to the level that the bicycle industry is used to.

EnergyBus will Raise the Overall Product Value

- EnergyBus will guarantee a certain level of vehicle functionality and quality.
- LEVs will become easier to use, to repair and to adapt to specific customer needs.
- Products will have a longer lifetime and require less maintenance.
- The customer will get more value for money.

EnergyBus will be a Key for LEVs to become a Mass Market

- With compatible components products can be adapted to specific customer wishes more easily; new applications will occur and product variety will grow.
- Product development will be faster, more flexible and cost-efficient.
- Lower production cost and more competition will bring the product price down.
- Compatible components, available spare parts, larger product variety and lower cost will boost the market to grow.

Applications

- This technology is very effective for development of LEV products. It is used in the manufacturing of LEV components as well.
- The concept is also applicable in the production of E-Bikes such as TDS (TourDeSuisse) E-Bikes, which is built on the Electragil system.
- It is also used in the public charge stations for Pedelecs which have been deployed in the Tegernsee Alpine region of Germany.
- This technology is also used in stationary systems. It includes the development of repeater stations in mobile communication network, off grid housing, measurement stations, off grid water pump stations and village DC grids etc.
- The Fraunhofer ISE also uses EnergyBus in their electric island systems located in Egypt. The system uses batteries and PVs, including the Kaco converters.
- The EnergyBus is used in the development of bike port system, motor home and boating equipment.

References

- Electric-vehicle-charging-station: techtarget.com, Retrieved 28 May 2018
- Technical-guide, electrification-transport: hydroquebec.com, Retrieved 18 April 2018
- Understanding-electric-vehicle-charging: pluginamerica.org, Retrieved 11 May 2018
- Electric-vehicle-charging-guide: fleetcarma.com, Retrieved 21 June 2018
- Types-of-connectors-for-electric-vehicles: wallboxok.com, Retrieved 31 March 2018
- Wireless-charging-electric-vehicles, automotive: powerelectronics.com, Retrieved 15 May 2018

Permissions

All chapters in this book are published with permission under the Creative Commons Attribution Share Alike License or equivalent. Every chapter published in this book has been scrutinized by our experts. Their significance has been extensively debated. The topics covered herein carry significant information for a comprehensive understanding. They may even be implemented as practical applications or may be referred to as a beginning point for further studies.

We would like to thank the editorial team for lending their expertise to make the book truly unique. They have played a crucial role in the development of this book. Without their invaluable contributions this book wouldn't have been possible. They have made vital efforts to compile up to date information on the varied aspects of this subject to make this book a valuable addition to the collection of many professionals and students.

This book was conceptualized with the vision of imparting up-to-date and integrated information in this field. To ensure the same, a matchless editorial board was set up. Every individual on the board went through rigorous rounds of assessment to prove their worth. After which they invested a large part of their time researching and compiling the most relevant data for our readers.

The editorial board has been involved in producing this book since its inception. They have spent rigorous hours researching and exploring the diverse topics which have resulted in the successful publishing of this book. They have passed on their knowledge of decades through this book. To expedite this challenging task, the publisher supported the team at every step. A small team of assistant editors was also appointed to further simplify the editing procedure and attain best results for the readers.

Apart from the editorial board, the designing team has also invested a significant amount of their time in understanding the subject and creating the most relevant covers. They scrutinized every image to scout for the most suitable representation of the subject and create an appropriate cover for the book.

The publishing team has been an ardent support to the editorial, designing and production team. Their endless efforts to recruit the best for this project, has resulted in the accomplishment of this book. They are a veteran in the field of academics and their pool of knowledge is as vast as their experience in printing. Their expertise and guidance has proved useful at every step. Their uncompromising quality standards have made this book an exceptional effort. Their encouragement from time to time has been an inspiration for everyone.

The publisher and the editorial board hope that this book will prove to be a valuable piece of knowledge for students, practitioners and scholars across the globe.

Index

R

S

T

www.ingramcontent.com/pod-product-compliance
Lightning Source LLC
LaVergne TN
LVHW080913210926
848485LV00004B/646